Civil Engineering Project Management

Civil Engineering Project Management

Fourth Edition

Alan C. Twort BSc, FICE, FCIWEM

and

J. Gordon Rees BSc(Eng), FICE, FCIArb

ELSEVIER
BUTTERWORTH
HEINEMANN

OXFORD AMSTERDAM BOSTON HEIDELBERG LONDON NEW YORK
PARIS SAN DIEGO SAN FRANCISCO SINGAPORE SYDNEY TOKYO

Elsevier Butterworth-Heinemann
Linacre House, Jordan Hill, Oxford OX2 8DP
200 Wheeler Road, Burlington, MA 01803

First published 1966
Second edition 1972. Reprinted in 1975, 1978, 1980, 1984
Third edition 1995
Fourth edition 2004

British Library Cataloguing in Publication Data
A catalogue record for this book is available from the British Library

Library of Congress Cataloguing in Publication Data
A catalogue record for this book is available from the Library of Congress

ISBN 0 7506 5731 6

Printed and bound in Great Britain

Contents

Preface

Most civil engineering construction projects are completed to time and budget but few get publicity for it. More often building projects are reported as exceeding time or budget because a building has to cater for the diverse needs of the many users of the building which can be difficult to forecast or may change as construction proceeds. In civil engineering the principal hazards come from the need to deal with below ground conditions, make structures out of re-assembled soils or rocks, and to cater for the forces of impounded or flowing water. The construction of roads, railways, tunnels, bridges, pipelines, dams, harbours, canals and river training measures, flood and sea defences, must all be tailored to the conditions found on site as construction proceeds because it is not possible to foresee such conditions in every detail beforehand.

As a result the successful management of a civil engineering project depends upon use of an appropriate contract for construction; the judgements of the civil engineer in charge and his team of engineering advisers; the need to arrange for supervision of the work of construction as it proceeds, and on the competence of the contractor engaged to build the works and his engineers and tradesmen.

The first four chapters of this book show the advantages and disadvantages of various ways in which a civil project can be commissioned, dependent upon the nature of the project and the needs of the project promoter. The recent legislative changes applying to construction contracts are noted, and the various different approaches now being adopted, such as partnering, 'PFI' and 'PPP' are explained and commented on. The book then sets out in practical detail all the measures and precautions the engineer in charge and his staff of engineers should take to ensure successful management and completion of a project.

The authors draw upon their experience in managing many projects both in the UK and overseas. Thus the book is intended to be a practical guide for project engineers, and a source of information for student civil engineers joining the profession. The author Alan Twort is a former consultant to Binnie & Partners responsible for many projects including the repair or reconstruction of several dams. Gordon Rees is a former Contracts Department Manager for Binnie & Partners and later Black & Veatch. He is now an independent consultant and an accredited adjudicator for ICE and FIDIC civil engineering contracts.

Alan C. Twort
J. Gordon Rees

Acknowledgements

Contributing author for Chapter 10 from Black & Veatch Consulting

E. Ruth Davies MSc, BEng, CEng, MICE, MIOSH
Safety Manager

Technical advisers from Black & Veatch Consulting

Keith Gardner CEng, FI Struct E, MICE
Chief Structural Engineer

John Petrie MSc, C Geol, FGS
Chief Engineering Geologist

1

The development of construction procedures

1.1 The nature of civil engineering work

Virtually all civil engineering structures are unique. They have to be designed for some specific purpose at some specific location before they can be constructed and put to use. Consequently the completion of any civil engineering project involves five stages of activity which comprise the following:

1. Defining the location and nature of the proposed works and the quality and magnitude of the service they are to provide.
2. Obtaining any powers and permissions necessary to construct the works.
3. Designing the works and estimating their probable cost.
4. Constructing the works.
5. Testing the works as constructed and putting them into operation.

There are inherent risks arising in this process because the design, and therefore the estimated cost of the works, is based on assumptions that may later have to be altered. The cost can be affected by the weather during construction and the nature of the ground or groundwater conditions encountered. Also the promoter may need to alter the works design to include the latest technical developments, or meet the latest changes in his requirements, so that he does not get works that are already out-of-date when completed. All these risks and unforeseen requirements that may have to be met can involve additional expenditure; so the problem that arises is – who is to shoulder such additional costs?

Clearly if the promoter of the project undertakes the design and construction of the works himself (or uses his own staff) he has to meet any extra cost arising and all the risks involved. But if, as in most cases, the promoter engages a civil engineering contractor to construct the works, the contract must set out which party to the contract is to bear the cost of which type of extra work required. The risks involved must also be identified and allocated to one or the other party.

1.2 The most widely used contracts for construction

One of the most frequently encountered risks in civil engineering construction is that the ground conditions met during construction will not be as expected, because trial boreholes and test pits cannot reveal the nature of every cubic metre below ground level. This means that quantities of excavation, filling, rock removal and concrete, etc., for such as the foundation of structures or laying of pipelines actually found necessary may differ from those estimated.

The risk that the promoter will need changes also arises from the relatively long time it takes, often 2 years or more, to get a civil engineering project designed and constructed. During this time it is always possible for newer processes or equipment to be developed which the promoter needs to incorporate in the works, or there may be revised forecasts of demand for the project output.

The traditional way of dealing with these risks of change is for the design of the works to be completed first, and then to produce a construction contract for which civil engineering contractors are invited to tender. The price bidders tender for such a contract is based on a *bill of quantities* which lists the estimated quantities of each type of work to be done, 'taken off' (i.e. measured) from the completed drawings of the works required. Against each item a contractor bids his price per unit quantity thereof, and these, multiplied by the estimated quantity of work to be done under each item, when totalled form 'the Contract Sum'. This system permits the contractor to be paid pro rata to the amount of work he actually does under each item, and also eases valuation of the payment due to the contractor for executing changes to the design of the works during construction to overcome some unforeseen difficulty or make an addition. The promoter can thus make reasonably small alterations or additions to the works required during the construction period – provided these are not so extensive as to 'change the nature of the contract'.

A standard form of contract using the 'bill-of-quantities method', was first introduced by the UK Institution of Civil Engineers in 1945. This standard form, known as the *ICE Conditions* became very widely used, and in the 7th edition is known as the 'Measurement Version'. A similar form of contract, known as the *FIDIC Conditions*, was developed by the International Federation of Consulting Engineers for worldwide use.

A basic provision of both these standard forms is that the contract between the promoter and the contractor for construction of the works, is administered by an independent third party – 'the Engineer' – who has the responsibility of seeing that the provisions of the contract are fairly applied to both promoter and contractor. The Engineer[1] has power to ensure the contractor's work is as the contract requires and issues certificates stating how much the promoter is obligated to pay under the terms of the contract. This avoided the bias that might occur if either the promoter or contractor decided these matters.

[1] 'The Engineer' (with a capital E) is used to distinguish the engineer appointed to administrate a contract for construction under ICE or FIDIC or similar conditions.

The great majority of all civil engineering projects undertaken by British engineers in the UK and elsewhere have been, and still are, constructed satisfactorily under the ICE or FIDIC Conditions. However, other methods are also commonly used to meet special requirements as shown below, and the ICE and FIDIC have developed other standard forms for such purposes (see Chapter 4).

1.3 Other long-standing procedures

Lump sum construction contracts

Under the standard ICE or FIDIC Conditions, the financial outcome of a project is not absolutely fixed, because the promoter has to pay for any extra work caused by conditions 'which an experienced contractor could not have foreseen'. This does not suit some promoters who wish to be certain what an intended project will cost, so 'fixed price' contracts came into use, often for a lump sum. Under them the construction contractor has to take all risks, such as meeting unexpected ground conditions. Such fixed price contracts can be satisfactory for both promoter and contractor for relatively simple, easily defined works involving little below-ground work.

Naturally a contractor's price for undertaking a contract for a fixed sum is higher than for a bill-of-quantities contract for the same work under which he is paid by measure of the work he is required to do. But this can suit a promoter who prefers to be certain about his financial commitment and where the works he requires can be well defined in advance. However, if the possible risks on the contractor appear high due to many imponderables – such as the works being large or complicated, or ground conditions being uncertain – then the extra charge made by the contractor for shouldering the risks may be high. Should the promoter require amendments as construction proceeds, then these will also prove expensive.

Cost reimbursement contracts

These contracts have been in use for many years on projects which involve unforeseeable amounts or kinds of work – such as the repair of a dam or collapsed tunnel, or repair of sea defences. Payment to the contractor is usually on the basis of: (i) direct costs of materials, labour and plant used on the site; plus (ii) a percentage addition for overhead costs; plus (iii) a fixed fee, or further percentage on for profit. Often a cost reimbursement contract for specialist work is negotiated with a suitably experienced contractor. If competitive bidding is required this would be based on comparison of contractors' quotations for overheads and profit. The advantage is that the promoter's engineer in charge of the project can work in partnership with the contractor to devise the cheapest means of overcoming problems. The main disadvantage for a promoter is that he carries all the risk of cost overruns, while the contractor is assured of his profit and fees. Where the works can be reasonably well defined, it may be best to use a measurement type of contract with a contingency sum allowed for any changes found necessary.

Sometimes a target cost is set under a cost reimbursement contract, the contractor sharing in any savings or excesses on the target cost. This gives the contractor an incentive to be efficient; but problems can arise if the target has to be altered because the work found necessary differs from that expected (see Section 3.1(e)).

Design and build contracts

These contracts are useful to a promoter who wishes to delegate the whole process of design and construction, or for whom gaining the output of a project is of more importance than the details of design. They also suit promoters who would not expect to be involved in construction work, such as health or education authorities. D&B contracts can offer a price advantage because the contractor can reduce his costs by using easy-to-construct, standard, or previously used designs which suit his usual methods of construction and existing plant.

A disadvantage to some promoters is that they lose control over the designs for which they are paying and may thus not get works wholly to their liking. Such contracts should only be used where there is little risk of the promoter's requirements changing during construction.

Since the contractor is taking on more risks including those of design and buildability, prices will usually be higher than for a measurement contract. Any attempt to achieve a short completion time for a project by use of such conditions may also lead to increased prices and possible overruns of time, as not all of the processes of design and construction can overlap.

1.4 Growing use of design, build and operate contracts

Design, build and operate (DBO) contracts were increasingly used in the 1980s onwards by government departments in the UK who saw a benefit in not shouldering all the complications of building and operating a new facility, but in passing this out to the commercial sector. Such contracts have the added advantage that if a contractor has to operate the works he has built for a number of years, he has a financial incentive to use good quality design and materials to minimize his expenditure on operation and maintenance.

There are several variations of DBO contracts. A BOT 'build, operate, transfer' contract usually implies the client pays for the works as they are constructed and takes over ownership of them at the end of the operation period. A BOOT 'build, own, operate and transfer' contract usually implies the contractor finances construction of the works (or negotiates with some funding agency to provide the funds) and transfers ownership of the project to the client at the end of the operational term of years.

A variety of ways of funding DBO contracts and re-imbursing the contractor are possible. Where a contractor provides all the finance required under a BOOT contract and receives income from the project output, this approach is indistinguishable from 'Private Finance Initiative' (PFI) described below – save that, under BOT and BOOT contracts the promoter usually identifies the size and nature of project required, whereas under PFI the contractor may do this.

1.5 Developments in the later 1980s

During the 1980s, as competition between civil engineering contractors for jobs in the UK intensified, contractors tended to reduce their margins for profit and risks in order to gain work. Consequently a contractor getting a job with low margins had to protect his position by making sure he billed the promoter for every matter he was entitled to charge for under the contract. However, some contractors developed the practice of submitting claims for extra payment wherever they thought a weakness in the wording of the contract might justify it. They employed quantity surveyors for this purpose, and it was not uncommon for more than a hundred claims of this type to be submitted on a major project.[2]

The resulting 'climate of dispute' that seemed to arise – more particularly on complex building projects than in civil engineering – led to other methods being sought for controlling constructional work. Some promoters thought that the independent Engineer, who had to decide on claims under the ICE or FIDIC conditions of contract, was not being tough enough in rejecting contractors' claims. But claims would inevitably arise and some have to be paid, especially in cases where a promoter did not allow enough time and money to be spent on site investigations, or who let construction start before being certain of his requirements. The practice of promoters to accept the lowest tendered price on most projects also increased the chance of employing a contractor whose price was so low he needed to use claims to safeguard his precarious financial position on that contract.[3]

[2] Before about 1975 most civil engineering contractors did not employ quantity surveyors. It was only the building industry which used them.

[3] While a commercial company can place a contract with any contractor it favours a public authority must 'safeguard the public purse', and cannot therefore reject the lowest tender without good reason. But, although an experienced engineer can see when a tender price is perhaps too low, he cannot *prove* this is bound to cause trouble. Nor can he guarantee that the next lowest tender, if adopted, will be free of trouble over claims.

In 1985, in an effort to reduce claims, the UK Department of Transport (DTp) proposed to deprive the independent Engineer of his role in settling claims under contracts for motorways and trunk roads, and let one of their own staff decide what should be paid. The DTp faced especial difficulties because road building involves much below-ground work and building in earth. Even minor changes in below-ground material from that expected can give rise to large extra costs for the contractor.[4] However, due to wide opposition, the DTp did not pursue its original intentions. Instead both the DTp and other public bodies sought to have more say in decisions on claims, such as giving the promoter a right to have his own staff take part in discussions with the contractor on claims, or requiring the Engineer to consult with the promoter on any claim exceeding a given amount.

1.6 New approaches to construction contracts in the 1990s

In 1991 the ICE published a new type of contract called the *New Engineering Contract* (NEC) which aimed to promote better management of construction contracts to reduce claims and disputes. Under it a Project Manager acting for the Employer administers the contract and a separate Adjudicator is appointed to settle disputes, subject to later arbitration or legal settlement (see Section 4.2(f)).

In 1994 changes to the construction industry to improve its efficiency were proposed by Sir Michael Latham in his report, *Constructing the Team*, commissioned jointly by the government and industry. He recommended use of standard contracts with payment and dispute terms defined, such as in the NEC, setting up registers of approved consultants and contractors for public work, and measures to protect contractors and subcontractors against delayed or non-payment. The more radical of his proposals met opposition and were not adopted; but his report resulted in the UK government passing the *Housing Grants, Construction and Regeneration Act 1996*, Part II of which, dealing with Construction Contracts, adopted a number of Sir Michael's key recommendations. This Part II required that all contracts for construction should provide for

- the right of a party to the contract to refer a dispute to adjudication;
- entitlement of a party to the contract to be paid in instalments;
- no withholding of payment due without prior notice;
- payment not to be made conditional upon the payer receiving payment from a third party unless the latter became insolvent.

[4] A technical paper published by the ICE (Paper No. 9999, 1992) showed how a reduction of 24 per cent in soil shear strength from that expected could result in an increased rolling resistance for earth moving equipment which reduced plant productivity by 37 per cent, costing the contractor that much more.

The Act came into force on 1 May 1998 and, where a contract did not include provisions required by the Act, *The Scheme for Construction Contracts (England and Wales) Regulations 1998* applied. This detailed an ajudicator's powers and duties; and the payment conditions required by the 1996 Act.

Most standard conditions of contract used by promoters to employ contractors already complied with the 1996 Act, but the Act also applied to contracts between a contractor and his subcontractors. The Act does not, however, apply to works for extraction of minerals, oil or gas; works for an occupier of a dwelling, or any works estimated to be completed within 45 days.

In 1995 the ICE produced a revised edition of the former 'New Engineering Contract' re-naming it the *Engineering and Construction Contract* (ECC), details of which are given in Section 4.2(f). Use of the ECC has increased steadily and now matches the traditional ICE forms. Both forms, when well managed, are capable of producing successful works with minimal disputes if the documents are carefully drawn up and the contract terms fairly applied.

1.7 Introduction of 'Private Finance Initiative'

In 1992 the UK government announced the introduction of the PFI for the procurement of infrastructure projects, such as roads, bridges, railways, hospitals, prisons, etc. Under PFI the whole cost of a project is met from private investment funds and the lenders of those funds look to the stream of cash flows from the earnings of the project for a repayment of (or a return on) their investments. The sponsors of a PFI project are usually a consortium of contractors and their funding banks who set up a company to undertake the project. The company receives loans from the sponsors (and often other banks) and may also raise equity capital, i.e. shares. It designs, constructs, finances, maintains and operates the project for a term of years under a concessionary agreement granted by the promoter who may be a government department, local authority or other public body.

An outstanding example of PFI was the Channel Tunnel. The initiators of the idea were two groups of banks and contractors – one British, the other French. After the English and French Governments agreed to support the project, the banks became the sponsors of it and set up the company *Eurotunnel* to fund, own and operate the tunnel under a 55-year concession from the two Governments. The contractors then joined together to form *Transmanche Link* to design and construct the tunnel. *Transmanche Link* was a holding company for two other executive companies, one an alliance of five British contractors to drive the tunnel from the English side, the other an alliance of five French contractors to work from the French side. *Eurotunnel* was, in effect 'the client' or promoter for whom *Transmanche Link* worked.

A PFI project takes much time and money to set up because of the long term of the contract and the many risks which have to identified and allocated to one or other of the parties. The contractor has also to negotiate with banks and other funding agencies for the necessary capital. Hence, only the largest contractors with substantial financial backing are able to undertake a PFI scheme.

The promoter has also to spend money on setting up an organization to check that the sponsors and their contractor comply with the terms of a concessionary agreement, and to resolve any problems occurring due to changed circumstances arising during construction and the period of the agreement.

1.8 Public–Private Partnerships

There are certain infrastructure and other public works which are not favoured for PFI because they do not give the assurance of providing an adequate return on funders' investment. Yet it may be in the interests of public authorities to involve a private contractor in executing a project because of his experience and efficiency (when well run) and the capital contribution made by the contractor and his funders which reduces capital borrowings by the public sector.[5]

Thus instead of PFI some form of Public–Private Partnership (PPP) may be adopted, under which the public authority takes on some risks in order to make the project attractive enough for the contractor and his financial backers to undertake it. Thus if a road is constructed and financed by a contractor and he is to be rewarded by 'shadow tolls' on the number of vehicles using the road annually, the promoter may guarantee a minimum payment to the contractor. Thus the public authority takes the risk of traffic being less or more than that estimated.

There are many other possible arrangements under PPP. Some PPP projects are 'quasi PFI' such as when a public authority provides a grant towards the capital cost, or arranges for a grant to be received from some other funding body, such as the European Community (EC).

1.9 Partnering

The *Egan Report* of 1998[6] had a wide effect on the construction industry because it suggested many ways in which knowledge of good practice in design and construction could be more widely disseminated to achieve increased efficiency and also reduce costs, accidents, defects and time for construction. It emphasized that there should be more use of partnering and alliancing. This was

[5] This reduces the PSBR (public sector borrowing requirement) i.e. the total amount of government debt which, expressed as proportion of the GDP (gross domestic product), is used as an indicator of a nation's economic health – just as a person's 'credit-worthiness' is undermined if he falls into too much debt relative to his income.

[6] 'Rethinking Construction' by Sir John Egan. Report of UK Government Construction Task Force.

supported by the DETR's[7] *Construction Best Practice Programme*; the Government National Audit Office's report on *Modernising Construction*, January 2001; and the Local Government Task Force's publication *Rethinking Construction: Implementation Guide*, August 2001, which gave over one hundred recommendations to local authorities for better practice in achieving construction.

Although there can be many forms of partnering, in construction it most often involves a promoter, his designers, and the contractor or contractors for construction. Although the 'partnership' need comprise only a statement of good intent by the parties, it can be more firmly established as a contractual relationship. Each of the contracts entered into by the promoter then contains a clause requiring co-operation with the other parties. Any of the usual forms of contract can be used, dependent on the nature of the work involved and provisions for payment. For some projects the partners may be required to keep their books open for inspection, or cost-reimbursement contracts can be used to provide the necessary information.

The partners and their staffs work together as a group to identify better methods of working and overcoming potential problems and to resolve these to the benefit of the project and the partners. Specific objectives may be set and incentives applied to encourage co-operation of the partners, perhaps by means of risk sharing and bonus or damages payments depending on the outcome of the project. It is important to recognize that the partners may change over the time scale of a project and that not all those involved in a project need to be partners.

Partnering may apply to long term alliances where the same teams may produce a series of works with the intent of improving the product and reducing costs. This could, for example, be for such as repeated roadworks, sewer or water main relaying, or even major works of a similar type. Alternatively partnering can be for single projects in which case project specific objectives may be set.

1.10 Project Management

Project Management became an 'in vogue' term in the mid 1990s, primarily to emphasize the need for management to be efficient to ensure successful completion of a project. However, the term covers many possible arrangements.

Companies were set up to provide management services. A promoter could, for instance, use a management contractor to manage the construction of a project under either of the arrangements termed *Construction Management* or *Management Contracting* described in Section 2.5. Alternatively where a promoter requires management of both design and construction of a project, one of the procedures described in Section 2.6 can be adopted.

[7] The Government Department of the Environment, Transport & the Regions.

Consulting engineers have, of course, always provided independent project management services to a promoter for the design and construction of a project.

This book deals with many of the arrangements for project management that are possible, commenting on their benefits and weaknesses, and detailing the practical measures which should be adopted to ensure the successful conclusion of a project.

However, the term 'Project Manager' has such a broad meaning that it is often used loosely to apply to people occupying quite different positions in various organizations. The following terms are more explicit so are used where necessary.

- *The Project Manager* is preferably confined to mean the person acting on behalf of the promoter to administer a contract for construction, as defined in the ICE's ECC contract conditions (see Sections 1.6 and 4.2(f));
- *The Agent* – a long-standing traditional term – is preferably used to designate the contractor's person in charge of construction on site;
- *The Project Engineer* can be used to designate the key executive person (usually an engineer but not always) delegated by a promoter or consulting engineer to be in charge of the design of a project – who usually also draws up the contracts for construction and sees the project through to completion.
- *The Resident Engineer* – another traditional term – can be used to designate the 'Engineer's Representative' on site to oversee construction, as defined in the ICE Conditions of Contract (see Section 9.1).

1.11 Operational or service contracts and 'Facilities Management'

Contracts to operate and maintain works not only form part of BOT, BOOT and PFI contracts mentioned in Sections 1.4 and 1.7 above, but are increasingly being adopted separately. Various terms apply having different shades of meaning. General terms are *Operational Agreements* and *Facilities Management (FM)*. If only certain operations are undertaken by a contractor this is termed *Contracting Out*. Various forms of *FM Contract* can be let under which a management contractor is employed to run, or advise and direct existing staff how to operate, a facility at maximum efficiency, or how to undertake construction of a new facility, tutoring staff in new techniques.

Leasing Agreements are somewhat different in that they comprise a contractor taking over and running the operation of some works for a period for a fee, or for part of the income from sales of the works output. Such an agreement can also require the contractor to maintain or refurbish plant and equipment, and introduce new equipment, so that the contractor has a financial input which the terms of the agreement need to cover. In many French cities, leasing – termed *Affermage* – has been widely used for many years for the operation of waterworks, or wastewater systems. The contracts are for a term of years. The advantage is that a contractor who specializes in such work has a wider range of specialist staff to draw upon than a small local authority.

Concessionary Agreements are those used where PFI is undertaken, as described in Section 1.7 above.

Facilities Management Contracts can cover the running of practically anything; from providing personnel to maintain a prison, school or sewerage system, to providing janitors and window cleaners. Such contracts are increasingly being undertaken by civil engineering contractors and even some consulting engineering firms, because of their management skills.

1.12 Framework Agreements

These are defined in EC Directives and UK Regulations for Utilities (SI 1996/2911) as:

> a contract or other arrangement which is not itself a supply or a works contract but which establishes the terms (in particular the terms as to price and, where appropriate, quantity) under which the provider will enter into such contract with a utility in the period during which the framework agreement applies.

The promoter defines a type of work for which he wishes to let a number of contracts. From an open tendering process a shortlist of firms are selected on the basis of some pre-set criteria such as experience, staff proposed, financial resources, etc. The promoter then invites the selected contractors to bid prices for future works of the kind defined. When the promoter requires some of the works defined he can then negotiate terms for it with a contractor on the basis of prices already submitted.

The advantage of framework agreements is that they avoid the need for open tendering, or repeated prequalification of tenderers under restricted or selective tendering. A criticism, however, of framework agreements is that they can result in long-term tie-ups between a contractor and an employer, thus tending to reduce open competition as mentioned below in Section 1.14. Also, as with any pre-selection of contractors on the basis of experience, etc. first, and then on competitive prices submitted, this could drive prices down unfairly.

1.13 Influence of computers and information technology

The use of computers, e-mail, and the Internet for transfer of designs and data has now become commonplace. Copies of drawings and documents produced in one office are now sent via the World Wide Web to other offices anywhere in the world, so that the recipients' comments thereon can be returned without delay. Thus specialist engineering guidance centred in one part of an organization can service the needs of others in distant locations. Computer aided drafting (CAD) has also become the norm for production of all formal drawings and to some extent for engineer's sketches.

A further development has been the integration of the design process with the requirements for construction. An **Intranet** can be set up to link people together within their own organization. Drawings and design information, specifications and bills of quantities for a construction project can be stored centrally and accessed by all authorized members of a design team, with only certain members authorized to alter the details. This means that drawings and data being used are always current, there are no delays caused by awaiting information, and the process of making changes can be controlled and audited. A design change can, for instance, lead to an immediate change in the contract drawings and the relevant specifications and bills of quantities.

A natural development has been to extend the availability of data to other parties concerned with a project – such as the employer, the principal contractor, and perhaps to certain specialist suppliers or advisers involved – by setting up an **Extranet** using the Internet. Such arrangements can be variously termed 'project collaboration' or 'project portal' systems. But greater care then has to be exercised in the selection of information made available on-line, in restricting access to it by only certain authorized parties, and in providing adequate security protection. This type of Extranet collaboration is also useful between firms when Partnering, Alliancing, or Joint Venture (see Sections 1.9 and 1.15) arrangements are adopted. An Extranet system is usually procured from a specialist website service provider, and has at least two main divisions – (i) a data division containing the basic information deemed necessary; and (ii) a division for recording inter-party communications. Computer software must be compatible and the set-up cost can be high, so that Extranets are mostly used for large projects.

Some difficulties can arise with computerized project collaboration. There is doubt whether a contract instruction from one party to another via such a system is legally valid in UK where contracts normally require instructions to be issued 'in writing'. There are also potential problems in preserving copyright of designs. Whereas 'hard copy' contract drawings provided to a contractor must usually be returned to the design engineer on completion of a contract, there is no equivalent precaution that can be taken when drawings can be archived on disks.

The fact that the various parties inter-connected can communicate freely with each other can also tend to blur responsibilities. Care must be taken to ensure that communications conform to the contractual position each party holds, so that misunderstandings do not arise. Also the ease with which key specialist advisers or project managers can be contacted can result in them being overloaded with requests to assent to some proposed action. The danger this creates is that, with limited time for the specialist or manager to consider all the ancillary circumstances applying to the often complex problems arising in civil engineering, incorrect or insufficient advice is given. The sending of copies of a communication 'for information only' to parties additional to the main intended recipient may also have to be restrained to prevent too many documents cluttering computer screens. The indexing of data files covering much diverse data also needs careful pre-planning to provide an adequate definition of the contents of each file and avoid mis-filing of further data added.

1.14 A criticism of certain systems

A criticism of systems such as BOOT and PFI is that only the larger contracting firms with large financial resources or sufficient financial backing can undertake them. Such systems also tend to utilize the services of a major contractor and his subcontractors for long periods. Consequently if the use of these forms of contract by promoters should become too widespread, there may be an insufficient number of large contractors left for proper open competition to occur for new projects, and a promoter may have difficulty in obtaining satisfactorily experienced bidders interested in a project he wishes to undertake. The smaller contractors may be forced out of business as promoters use these systems or favour work packages too large for the smaller contractors to undertake. Also the best quality subcontractors and suppliers can become tied to one or another major contractor for long periods, and not be available to serve other contractors. The end result could reduce competition on price and quality between contractors which is still of importance in fostering the development of innovative methods and improvements in efficiency.

An objection frequently voiced is that PFI projects must be more expensive than publicly funded projects, because the shareholders and commercial lenders financing PFI want a higher return on capital than is paid on loans raised by a public authority. However, shareholders usually provide only a small proportion of the capital required for a major project[8] the rest being provided by loans from banks and other financial organizations on which the interest charges are only a little above the interest charges payable on public loans. Hence, the overall cost difference between private and public funding can be relatively modest. However precise evaluation of the cost differential is complicated because account has also to be taken of such matters as the administrative costs in setting up PFI, the different sums involved to cover risks, project maintenance and supervision thereof over a long term of years. One estimate suggests the cost of private finance is about 3 per cent higher than public finance.[9] A cost difference of this order does not seem particularly

[8] In the case of the Channel Tunnel, 25 per cent of the capital required was in shares, 75 per cent being in the form of loans. For the proposed modernization of London's Tube lines, a consortium of bidding contractors are reported as aiming to provide £180 million share capital and raising £2000 million from bank loans.

[9] Grubb S.R.T. 'The private finance initiative – public private partnerships' *Civil Engineering*, August 1998, pp. 133–140.

significant. A more significant factor in reducing project cost is the efficiency with which promoters, designers and contractors carry out their roles.

1.15 Ancillary contractual practices

By the end of the 1990s the construction industry had tried out a variety of permutations of construction procedures, most of them being only ancillary practices attached to one or other of the main approaches already described above. The following list gives the new terms most frequently used and their meaning. Few are radically new practices, and some had a phase of popularity which has already declined.

Alliancing A term principally applying to a contractor who joins with one or more other contractors to undertake a contract for some project. One firm is the lead firm; the others are often specialists. An example is an EPC Contract (Engineer, Procure, Construct Contract) under which a firm of consulting engineers may be the lead firm (see Section 2.6(c)) with a construction contractor and plant suppliers associated. Other setups are possible, such as when a construction contractor or plant supplier is the lead firm and uses consulting engineers to design the structures required. Alliancing is also sometimes used as an alternative name for Partnering.

Benchmarking A procedure under which a promoter (or manufacturer or contractor) compares his performance achievements on projects with the methods and achievements on similar projects carried out earlier by him, or carried out by some other promoter. It involves comparing such things as project cost per unit of some kind; time and cost over-runs against that intended; disputes and troubles encountered, etc.

Best Value Contracts The requirements placed by government on UK local authorities in place of Compulsory Competitive Tendering (CCT) (see below). Tenders for construction or provision of services now have to be chosen not only on bid price, but also on the quality of the materials and services offered, as affecting the estimated operational and maintenance costs of a project and its estimated length of life. This is evaluating bids on a ' whole life costing' basis.

Competitive dialogue Pre-bid negotiations initiated by a promoter who, not having defined his project requirements in any detail, invites outline proposals from contractors for a design and build project as part of the prequalification stage for prospective bidders. Criticisms of the procedure are that the promoter gets useful advice on design alternatives without paying a proper design fee for same, and that the promoter may choose the best design submitted by one contractor but use another contractor to execute it.

Compulsory Competitive Tendering The procedure that the UK government previously required local authorities to adopt, before they introduced

Best Value Contracts (see above). It meant that in-house local authority staff had to compete on price against contractors' bids for constructing a project or providing services.

Construction Best Practice Recommendations of the UK 'Construction Best Practice Programme' to promoters and contractors for improving productivity and efficiency, following the findings of the 1998 Egan Report, *Rethinking Construction* (see Section 1.9 above).

EC Procurement Regulations EC rules have for some years required open competition for certain types of work, as set down in EC Public Procurement Directives. These require that all public utilities and other major public organizations put tenders for services and construction out to a tender system open to all EU firms. The rules apply for values of projects above certain minimum figures and require that details of contracts open to tender are published in the Official Journal of the European Community. Contracts must be tendered individually or for groups of contracts for specified similar types of work. Individual tendering can lead to large numbers of bids being received, each requiring analysis; whereas grouping allows a short list of preferred bidders to be developed. Further details of EC tendering requirements are set out in Sections 6.2 and 6.3.

The Gateway Process A system of adopting checks on the progress of a project at critical stages i.e. 'gates'.

Joint Ventures A relationship usually with a legally binding agreement in which two or more firms agree to combine resources to carry out a contract. The joint venture may be for consultancy or construction work. Between themselves the parties to the joint venture may divide up the work and decide on profit split or liabilities, but the main contract with the promoter will usually hold them jointly and severally liable for the outcome of the contract.

'KPI' or Key Performance Indicators These indicators are used for comparative purposes, are measures of success in the design and construction of a project. Chief measures are outcome cost as compared with estimate, time over-runs, promoter satisfaction, freedom from defects, and safety record.

'M4i' (Movement for Innovation) Promotion of new techniques by members of the 'Construction Round Table' in line with recommendations of the Egan Report, the members of the Round Table comprising representatives of a number of companies making large investments in new constructions.

One Stop Shop A colloquialism for the case where one contractor delivers all that is required to design and construct a project or series of projects.

Prime Contracting A form of design and build contract under which the prime contractor has an association with a number of subsidiary firms whom he uses to supply specialist goods or services. Thus, instead of a designer specifying nominated subcontractors to be used by the contractor, these are chosen by the prime contractor. This avoids the problems that can arise with nominated subcontractors (see Section 15.8).

Quality Assurance (or QA) This is defined as all those planned and systematic actions necessary to provide confidence that a product or service will satisfy given requirements for quality. Thus QA is concerned to ensure that adequate systems are set up for checking that work is properly done, and that such systems are complied with in practice. It is not a system for providing 'best' or indeed any specific quality of materials and workmanship, but only to ensure that adequate administrative procedures are adopted to see the specified requirements are met. All aspects of the construction process may use QA. Consultants, contractors or suppliers can set up QA procedures covering the whole range of work they have to do, including checking work done against those procedures and arranging for audits to demonstrate compliance. Permanent QA systems may be certified and audited by an independent organization or audits may be required by a promoter.

QA has the advantage of requiring people to manage their processes better, but should not be taken as eliminating the need for checking the methods and details of working. Checking that a procedure has been followed does not necessarily mean that the work has been done correctly. Hence, although a construction contractor may run a QA system, it is still necessary for a promoter to be satisfied that the works have been constructed properly; for which independent site supervision of construction on behalf of the promoter is the best assurance.

Value engineering A non-specific term applied to any exercise to find out possible savings, economies and better 'value for money' by investigating alternative designs, construction processes, ways of planning and meeting risks, etc. for a proposed project. The exercise often takes the form of arranging a special 'workshop study' in which the client, designer, and contractor and other parties involved take part, and put forward suggestions for discussion and investigation.

2

Procedures for design and construction

2.1 Promoter's obligations

Before a promoter can start on a civil engineering project it will be necessary to undertake a number of studies. These may comprise:

- **market demand studies** to define what are the needs the proposed project should meet, such as the size and quality of the project output or benefit;
- **economic and financial studies** to decide for how long a period it is economic for the project to cater for the foreseeable demand, taking into account the cost involved and how the project is to be financed;
- **feasibility studies** to ensure the project is engineeringly practicable, confirm its probable cost, and decide what methods should be adopted for the design and construction;
- **legal studies** to ascertain what statutory or other powers must be obtained to construct the project, including environmental approvals.

These studies are all interconnected. For the market and financial studies, the promoter may appoint economic advisers because a major problem to be resolved is how large should the project be (in terms of output or capacity) and whether it would be economic to phase the construction in stages.

The feasibility studies will need to investigate different options for providing the output, to ascertain how practicable it is to adopt phased construction, and what difference this would cause to capital outlays and their timing.

Legal advice will be necessary to obtain powers to purchase land, gain access, alter public rights of way, abstract water, discharge waste, gain planning approval, and meet environmental and other objections. A number of outside bodies may have to be consulted on these matters.

Special procedures, including presenting the case for a project before a public inquiry or gaining parliamentary approval can be necessary for many types of work. For projects internationally funded it will be necessary to meet the extensive requirements of funding agencies, such as the World Bank, Asian Development Bank, United Nations Fund, or European Community regulations. These requirements are often complex and may necessitate the employment of a firm of consulting engineers experienced in such work, together with financial and economic advisers.

It can take 2 years or more on a major project to conduct all the studies required and negotiate the powers required for construction. Even on a small project these matters can seldom be completed in less than a year.

2.2 Importance of feasibility studies

Feasibility studies of an engineering nature are needed for most construction projects. It can be an advantage to a promoter if he employs an independent consulting engineer to check the technical feasibility and cost of the project. The consulting engineer should be able to bring extensive design and construction experience in the type of work the promoter needs, and be able to offer economic solutions to problems his experience tells him are likely to arise. The value of an independent consultant is that he uses only his professional judgement in deciding what will serve the promoter's interests best. Such a consultant should have no relationship with any commercial or other firm which could have an interest in favouring any particular kind of development.

In the initial stages, the studies usually concentrate on various options for the location, design and layout of the project. The studies may include different methods for producing the required results, such as the alternatives of building a dam or river intake or sinking boreholes or buying water from an adjacent company to produce a new supply of water. Accompanying this work there will be data gathering and analysis, followed by the development and costing of alternative layouts and designs, so the promoter can be assisted to choose the scheme which seems most suitable.

Site investigations are particularly important, and sufficient time and money should be spent on them. Although they cannot reveal everything below ground, inadequate site investigations are one of the most widespread causes of construction costs greatly exceeding the estimate. On large or specialist projects, trial constructions or pilot plant studies may be necessary such as, trial construction of earthworks, sinking of test borings for water, setting up pilot plant to investigate intended process plant, or commissioning model tests of hydraulic structures.

The feasibility studies should include a close examination of the data on which the need for the project is based. Many instances could be quoted where large sums of money have been saved on a project by carrying out, at an early stage, a critical examination of the basic data the promoter has relied upon. This data has to be tested for accuracy, reliability, and correctness of interpretation.

2.3 Options for design

The following shows the principal design options commonly adopted.

(a) Design by promoter or a consultant

The whole of the design, including all drawings and specifications, is completed before construction tenders are sought – except for drawings not needed for tendering purposes, such as for concrete reinforcement.

A promoter may have sufficient staff to undertake design work 'in house' or he may put all design out to a consultant, or divide the design work between them.

On schemes involving different types of engineering, design may be let out in separate 'packages' to different specialist consultants. For instance the design of an industrial estate may be packaged into – roads and drainage; water supply and sewerage; power supplies, and landscaping. For large schemes the promoter may appoint an overall consultant with wide experience to co-ordinate the inputs of the specialist design consultants.

Some elements of design may be left for the construction contractor or his sub-contractors to undertake, such as the design of heating and ventilating systems, or the cladding for a building. Specialist suppliers may need to design their product or services to suit the project.

Advantages are:

- The promoter can check all aspects of the design to ensure they meet his requirements before construction starts.
- Competitive tenders for construction are obtained on a clearly defined basis encouraging construction contractors to submit lowest prices.
- The risk of having to make alterations to the work during construction is minimized, giving a better chance of the project cost not exceeding the tendered price.
- The promoter is not committed to proceed with construction until he sees tendered prices and accepts a tender.

(b) Outline designs provided with detailed design by others

The promoter draws up outline designs and a specification of his requirements. He appoints a firm or firms of specialist designers to carry out detailed design, and then engages a management contractor to co-ordinate both the detailed design and the construction. This type of arrangement can be seen in some management contracts (see Section 2.5).

(c) Layout design by promoter; detailed design by contractor

The promoter specifies functions and design standards, and supplies layout plans. The contractor then undertakes the detailed design before proceeding with construction. The works may be relatively small, such as the design of a retaining wall; or fairly extensive such as the design of an intake and drainage pumping station, or the structural and reinforced concrete design for a water tower.

Advantages are:

- The contractor can adopt designs suiting his constructional equipment and his usual construction methods, enabling him to tender his lowest price.
- The cost of making design alterations during construction do not fall on the promoter.

Disadvantages are:

- The design may tend to suit the contractor more than the promoter.
- Control over design details is lost to the promoter.
- The contractor must increase his price to cover design risks.

(d) Functional specification by promoter: design by contractor

The promoter specifies the functions the project is to perform, for example the size, quality and performance criteria for the intended works. He also provides drawings showing the location of the intended works and draft layouts for them, and may specify standards for design. The works required may be extensive, such as design of a road, or the civil works and plant for sewage treatment works. The contractor undertakes the layout and detailed designs to the standards required.

This is the basic set-up for design and build (D&B) contracts where most of the design responsibility is held by the contractor. The advantages, disadvantages and complexities of such contracts are dealt with in Section 2.6.

2.4 Options for construction

(a) Direct labour construction

The promoter uses his own workforce to carry out construction. This gives the promoter full control of the work and flexibility to alter it. However, with no competition on prices, costs can be high unless management of the work is efficient.

Direct labour construction was common for works in Britain and for all sizes of projects overseas until the 1950s. It has continued overseas where sufficiently experienced local contractors are not available. Local authorities and public utilities in the UK continued to use direct labour for such as re-surfacing roads, constructing minor roads, laying water mains or sewers, etc. until the 1980s when the government required such jobs be opened to competition from contractors (see 'Compulsory Competitive Tendering' in Section 1.15).

Direct labour construction can be undertaken by consulting engineers on behalf of the promoter. The consultants hire the necessary labour and plant, and order the necessary materials, using money provided by the promoter. This procedure was widely adopted up to the 1950s for projects in the UK and overseas, and can still be used now. It was used on some works for raising the Essex side of the Thames tidal defences 1974–1984. Given a small team of engineers and some skilled foremen to guide local labour under a resident engineer with strong managerial capacities, direct labour under the control of a consulting engineer has often been notably successful in keeping a project to time and budget.

(b) Construction divided into trades

A practice often followed in developing countries is to split construction work into packages by trade, for example, brickwork, carpentry, etc. because local contractors often provide only one type of trade work. 'Self-build' houses in the UK often use this approach. The same approach on a larger scale is sometimes adopted for complex building projects, with a management contractor appointed to co-ordinate the work (see Section 2.5(b)).

(c) Main civil contractor supplies all ancillary services

Most civil engineering works incorporate services of an electrical or mechanical kind, such as for heating, lighting, ventilation and plumbing. It is usual to permit the contractor to choose the sub-contractors who provide such services, subject to the approval of the promoter. The promoter, however, must make provision in the design to accommodate such services.

An advantage to the promoter is that co-ordination of the sub-contractors then rests with the contractor, and if they delay him, that is his responsibility.

A disadvantage is that if the promoter specifies (i.e. 'nominates') some particular supplier of services or goods, the promoter then becomes responsible for any delay caused to the civil contractor by the nominated firm.

(d) Civil contractor constructs; promoter orders plant separately

When major plant such as generating plant, pumps, motors, or process plant has to be incorporated in civil engineering works, there is an advantage in the promoter letting separate contracts for such plant. This may be essential in cases where plant is on such long delivery time that it must be ordered before the construction contract is let. A discussion of the measures necessary to co-ordinate the plant contracts with the construction contract is given in Section 5.6.

Advantages to the promoter are that he has direct access to the plant supplier to specify his requirements and agree all technical details. He can receive plant drawings in good time to complete the structural designs.

A disadvantage is that, if the plant supplier is late on his promised delivery, the promoter may have to pay the contractor for delay. To guard against this, plant delivery times quoted to the civil contractor can allow a 'safety margin' on the plant supplier's quoted delivery time. The majority of all projects incorporating major plant are managed satisfactorily on this basis.

(e) Civil contractor orders all plant

On a large and complex project there may be an advantage in requiring the civil contractor to order plant, as specified and pre-agreed by the promoter with the plant supplier – provided the time for construction is long enough for plant to be delivered in time.

Advantages are:

- The civil contractor can be left to arrange delivery of pieces of plant to suit his construction programme.
- The civil contractor has direct contact with the plant supplier to agree to the details of any storage or lifting requirements.
- The promoter avoids the risk of delaying the contractor by not getting the plant supplier to deliver in time.

Disadvantages are:

- The plant supplier will not start manufacture until the civil contractor places his order.
- To complete the civil works design, the promoter may have to pay the plant supplier a fee for providing layout drawings in advance.
- If the promoter asks for some alteration to the plant, or a 'works test' on the plant shows the need for some amendment, delivery may be delayed causing the civil contractor to claim for delay.
- The plant supplier may increase his charges if he thinks his risks will be increased by having to rely on the civil contractor for payment.

(f) Plant supplier arranges building design and construction

Where the supplier of process plant exerts a dominating influence on the design of a project, the promoter may ask him to employ a civil engineering contractor as sub-contractor to construct the works to accommodate the plant. The plant supplier may then use some firm to design the civil works, or else he passes this also to the civil contractor.

Some plant suppliers, however, will not agree to this procedure, on the basis that either they have no experience of construction work or do not wish to be involved in it.

2.5 Construction using forms of management contracting

An alternative to the promoter or his consultant drawing up and letting contracts for construction of a project, is for the promoter to use a 'management contractor' to do this. There are two main forms of management contracting.

(a) Construction management

This term is used to mean the arrangement under which the promoter appoints a manager with his own staff to organize the letting and supervision of construction contracts which are placed by the promoter. Design may be by the promoter's staff, or can be placed as a separate design package or packages let by the promoter, but supervised by the manager.

An advantage is that an experienced construction manager should be able to avoid or minimize the problems of co-ordinating contractors. Disadvantages include the separation of the promoter's design requirements from construction supervision, and the extra cost of the manager and his staff.

(b) Management contracting

This is an arrangement more commonly adopted for complex building constructions rather than for civil engineering works. Under it the promoter appoints one contractor to manage all the construction inputs by letting contracts himself. These 'works contracts' are effectively sub-contracts to the management contractor. Many may be labour-only contracts, while others are for 'supply and erect'. The promoter may retain rights to approve or disapprove appointment of a works contractor. The promoter may also let a separate design contract, which is placed under the administrative charge of the management contractor.

Advantages are that the promoter is relieved of the responsibility for the letting of the many sub-contracts used, and the co-ordination of their inputs to meet the design required.

Disadvantages are that the speed of construction depends upon the ability of the management contractor to get efficient sub-contractors working for him. Some projects have been highly successful; others have suffered disastrous delays. Also, if construction starts before designs are sufficiently complete, any design alterations found necessary later can result in delays and excessive cost over-runs. A tangle of legal claims and counter-claims can then arise as each of the parties involved – the promoter, management contractor, designers, and works contractors – tries to make others responsible for some or all of the cost over-run. The price risk to the promoter is relatively high, since the terms of a typical management contract permit extra costs and risks to pass straight through to him from the works contractors.

Management contracting was initially much favoured for large building developments with associated civil engineering work; but there has been considerable debate concerning its merits and the number of jobs using the method has declined.

2.6 Design and build procedures and other options

(a) Design and build or 'turn-key' contracts

Contracts of this type are often for a lump sum which can suit a promoter who wants certainty of price, and who can be given a clear idea of what he is being offered. For instance, the contractor may be able to offer an 'off the shelf' design for a type of structure he has previously built and can show the promoter. Where this is not the case, the promoter may provide a drawing of what he requires and stipulate design requirements, for example, design processes and parameters to be used.

Advantages are:

- The promoter does not have to employ a separate designer.
- Construction can start before designs are complete and any consequent changes found necessary are the contractor's responsibility.
- Control of the design process permits the contractor to keep costs as low as possible by such measures as – using parts of previous designs, minimizing the need for complicated formwork, and tailoring dimensions to suit the contractor's equipment.
- For uncomplicated or traditional civil engineering work, or repeat structures of a kind the contractor has done before, a turn-key contract can give a promoter a satisfactory job at lowest price.
- There is also a possible advantage that collaboration between design and construction staffs can foster innovative design which reduces costs. But if the tender period gives insufficient time for an innovative design to be fully worked out, the contractor may think it too risky to allow for it in his tender. If later, the innovation proves possible, the contractor benefits and not the promoter.

Disadvantages are:

- If the design has yet to be formulated, the promoter has to leave most details in the contractor's hands.
- If the promoter employs a consultant to check the contractor's design, he will only be able to insist on compliance with matters specified in the contract.

- The promoter may need to employ an inspector to watch the contractor's construction.
- Bidding costs for other than simple structures are expensive, so contractors may refuse to bid if more than three or four are invited to tender.

If the promoter does not employ a consultant or inspector to check the contractor's work, his only real control over its quality and the end result is his checking of the packages offered by tenderers before awarding the contract. This is not necessarily sufficient because, in the limited time available for tendering, the contractor cannot work out all the details of his design nor specify the exact nature of everything he will supply. Thus the promoter can suffer disappointment at what he receives; and if he then wishes to make any changes these may be very costly or even impracticable.

(b) Design, build and operate contracts

Under this type of contract the contractor is required to operate and maintain the works for a period of perhaps 3–5 years after he has completed their construction. The contract may be for a lump sum, a proportion of which is payable in stages during the operating period, or income may be derived from sales or charges – bridge tolls for example.

Advantages are:

- The contractor is given an incentive to design and construct well, in order to ensure low maintenance and repair costs during the operating period. This is useful to a promoter who, for instance, wants a road built, because problems arising from faulty design or construction tend not to be revealed except under two or three years' trafficking.
- The operation provision reduces the promoter's need to check the contractor's work.
- The maintenance provision keeps the contractor available to undertake repairs during the operating period, though the promoter must have powers to act if the contractor does not undertake repairs and maintenance properly.

Disadvantages are:

- The same as those listed for design and build contracts under (a) above.
- The contractor has to shoulder added risks so his price can be high.
- The contractor's costs of bidding are higher than for a D&B contract.

A problem is that repairs or excessive maintenance could arise from unforeseeable ground conditions or, in the case of a road for instance, from traffic loading exceeding that specified in the contract, so occasions for dispute could arise. The promoter will also be responsible for any repairs due to an inadequacy in his specifications for design and construction. Where design, build and operate (DBO) contracts are for provision of buildings and process plant, such as for water or wastewater treatment, it is the quality of the equipment and consequent output which is principally tested by the period of operation.

Thus problems can occur if faulty performance is partly due to conditions arising which are not covered by the promoter's specification.

Where a DBO contract is let on the basis that the contractor also finances the project, associating with a bank for the provision of the necessary funds, the operating period may then be long term, for 15–20 years or more. This is typically a Private Finance Initiative (PFI) project, described in Section 1.7. The risks on the contractor are then increased since they include a substantial dependency on the terms of the income he is to receive. The promoter has the cost of setting up a long-term supervisory system to cover the operation period and may face the risk of circumstances arising which are not covered in the original contract.

(c) Engineer, procure and construct contracts

An engineer, procure and construct (EPC) contract is a form of D&B contract under which a design engineer or firm of design consultants heads a team which includes an experienced contractor and perhaps a plant supplier. The promoter specifies his project requirements in outline which the team designs in detail in continued liaison with him. The EPC organization arranges and manages construction, letting specialist work packages out as necessary to suitable sub-contractors. The promoter pays the actual cost of the work plus a fee, subject to a guaranteed maximum price, or to a target cost with an arrangement for the sharing of savings or excess costs on the target.

(d) Partnering

Details of partnering are given in Section 1.9. There are two types: 'term (or full) partnering' which covers an intention to carry out a series of projects together or for a given period; and 'project-specific partnering', i.e. co-operation for one job at a time.

Normally a promoter negotiates a partnering agreement with his consultant (if he employs one) and a contractor of his own choosing, usually because of past satisfactory experience of working with him. If competitive tendering is required, then a selected list of contractors may be invited to bid – on the basis of experience, quality of staff available, and costs plus charges for overheads and profit, etc. (similar to cost reimbursement contracts outlined in Section 1.3). But if open competitive tendering is used, the advantage of basing a partnering agreement on past successful working with a contractor may be lost.

(e) 'Term' or 'Serial' contracting

This comprises letting an ordinary construction contract for carrying out a series of works of an identical nature – re-surfacing roads, for example – for a given period of a year or longer. The terms of the contract set payment and other conditions for a series of similar works which are ordered from time to time as they are needed.

2.7 Comment on possible arrangements

A comparison between the two most commonly used methods of developing a project – completing design first and letting a contract for construction, or adopting a D&B contract – are given in Table 2.1. The advantages and

Table 2.1

Comparison of different methods of project promotion

Advantages to promoter	*Disadvantages to promoter*
Separate design, followed by contract for construction	
Designer chosen to suit nature of works	Promoter has to employ designer and check designs
Works designed suit promoter's requirements	Promoter may have to let separate contracts for special equipment and ensure delivery on time
Promoter can check designs before construction contract is let	Construction cannot start until designs are sufficiently complete
Competitive tenders for construction obtained on clearly defined basis	Contractor's experience does not contribute to design
Promoter not committed to project until construction contract entered	Promoter needs to employ engineering staff to supervise construction
Direct control over quality of construction work possible	Cost outcome dependent upon extent of unforeseen conditions met
Minor variations and additions to works possible during construction if ICE or FIDIC conditions used	
Design and construct combined contract	
Promoter needs no knowledge of construction	Extensive specification of requirements is needed
Promoter need not check design if he only requires guaranteed works performance	An engineer is needed to draw up specification
Promoter does not have to organize separate contracts for special equipment	Design unspecified may suit contractor and not promoter, because contractor designs for only the minimum necessary
Contractor can achieve low price by choosing designs cheapest to construct	Time for specifying, tendering, and examining tenders lengthy
Lump sum project cost known before construction starts	Tender comparison difficult because designs and materials offered differ
Construction may start before design is complete providing early completion	Promoter committed to works before they are designed
If project is simple a satisfactory result can be obtained at lowest practicable price	Changes difficult to make after contract is let and may be costly
	Promoter may need an engineer to watch construction
	Contractor's price may be high due to risks
	Number of contractors tendering may be limited

disadvantages listed depend very much on the size and nature of project required. Also, the choice between the two systems shown, and between them and all the other procedures described in this chapter depend on the promoter's resources, nature of business, and the restraints imposed on him by such as the need to conform with government rules, EC Directives, or his financial backers' requirements. In general, however, most types of arrangement can work satisfactorily if the contractor's prices are adequate, he is efficient and treated fairly, and the promoter specifies clearly what he wants and does not indulge in over-many changes.

The promoter who is able to plan well in advance so that he can define exactly what he wants and can give his designers adequate time to complete their work, will usually get best value for money. A contractor who tenders for works that have been designed in all essentials and which are not subsequently altered, will usually be able to give a good price and fast construction. Time spent ensuring adequate site investigations, full working out of the best designs, and careful production of contract documents, is the best guarantee that construction of a project will be trouble-free, on time and to budget.

3

Payment arrangements, risks and project cost estimating

3.1 Methods of payment under different types of contract

(a) Rates only contracts

These contracts call for tenderers to quote only their rates per unit of work of different kinds. They are used for work whose quantity cannot be defined in advance, such as for site investigations, grouting work or the sinking of boreholes. Any quantities entered in such contracts will be for indicating the amount of work expected and do not form a basis of the contract.

The tenderer has to ensure that his rates for each item of work carry enough oncost to pay for his overheads and profit. However, the items listed can include 'lump sum' prices for 'one off' costs, such as 'For bringing and setting up grouting plant on site' and so on.

In some overseas countries the government, local government authority, or public utility may publish its own standard rates for a range of civil engineering operations. Many of these will be for the provision of labour only, since pipes, steelwork and steel reinforcement are often supplied by the authority. Tenderers bid a percentage of the employer's standard rates to be applied to the quantities of work set out. Due to inflation and failure to update the standard rates, the percentages tenderers quote are often well over 100 per cent addition.

(b) Rates and prices for re-measurement contracts

These apply where ICE or FIDIC (or similar) measurement contracts are used, incorporating a bill of quantities for pricing, as described in Chapter 1

Table 3.1

Comparison of different methods of payment

Advantages to employer	*Disadvantages to employer*
– Re-measurement –	
Greater certainty in pricing	Detailed bills of quantities needed
Project cost known with reasonable certainty	Employer must meet extra costs due to changes
Facilitates valuation of variations	Work done has to be measured
– Lump sum –	
Firm price known early	Bid risks high hence tendered prices may be high
Financial bids easily evaluated	Long tender time required
Payment terms simple	Variations difficult to value and agree
– Reimbursable with fixed fee –	
Reduced documentation for tender	Bid evaluation difficult
Short tender time	Price competition minimal
Variations easily evaluated	Checking and auditing contractor's costs necessary
Cost of project can be controlled	No firm final cost
Methods of construction can be controlled by designer	Contractor has no financial incentive to minimize costs
Design can be altered during construction	Risk of inefficient contractor
– Reimbursable with target cost – *(as for reimbursable with fixed fee with following changes)*	
Contractor has financial motive to be efficient	Target difficult to define initially
Contractor shoulders some proportion of cost exceeding target	Target may need changing owing to changes in work required
	Disputes possible about fair target

Section 1.2. Where additional work is required, the contractor is paid for it at 'bill rates' or, if different work is required, this is paid at similar or agreed rates. Table 3.1 compares this method with the methods which follow.

The advantages are that the contractor can be paid fairly for the amount of work he has to do, and the employer only has to pay for work actually required, without having to pay a premium to the contractor for the risk of undertaking, at his own cost, extra work due to quantity changes. Thus if no major unforeseen conditions are encountered and the employer orders no extra work, the cost of the job to the employer will come very near the original sum tendered.

The use of bills of quantities has been the normal method of payment in standard forms of contract for many years. This method is particularly effective where the employer wishes to control the design, or has the works largely designed before going out to tender. With the works clearly defined, and a fair system of measurement, the contractor's risks are reduced and pricing may be keen.

(c) Lump sum contracts

A lump sum price may be called for, or a series of lump sums. This is best suited to easily defined, relatively simple constructions, involving little below-ground work. However, some quite large above-ground constructions are paid for by lump sum. Sometimes a separate section of the bill for pricing allows for the foundation work of a building to be paid for 'on measure'. In some kinds of civil engineering work the lump sum payment method can pose serious risks upon a contractor, causing him to add a substantial sum to his tender. This is particularly so for design-and-build or 'turn-key' projects where the contractor has to undertake detailed design as well as construction. The employer has to pay these additional sums whether or not any risks materialize.

A disadvantage is that an employer may have to pay a high price for any alteration or addition he wants to the project, because the contractor is only committed to undertaking a fixed amount of work for the fixed payment. Payments under lump sum contracts are usually made in instalments as set out in the contract according to stipulated stages of completion, or linked to a programme or activity schedule.

(d) Cost reimbursement contracts

Under a reimbursable contract the contractor is usually reimbursed his expenditure monthly on submission of his accounts, which must include evidence of payments made to suppliers of materials, gross wages paid to employees, and hours operated by plant. The invoices for materials have to be checked to ensure they are materials used on site. Plant rates have to be pre-agreed, and different rates may apply for plant 'standing' or 'working'; with lump sums payable for bringing plant to site and taking it away. Where a fixed fee has been agreed for his overheads and profit, this is usually paid in stages as the contract sets out.

Chapter 1, Section 1.3 describes the advantages and disadvantages of cost reimbursement contracts which are normally adopted only for work whose nature or extent is not defined in advance. The contractor usually remains responsible for constructing the works, his methods and expenditure being agreed with the employer, or on a day-to-day basis with the employer's project manager or resident engineer on site. An employer may be reluctant to adopt a cost reimbursement arrangement because the cost outcome is uncertain and the employer has to rely on the contractor to be efficient and not waste money. These objections can be overcome to some extent by adopting a target cost approach as described below.

Under any cost reimbursement contract it is essential to detail just what costs are to be paid, and which are covered by the fees or other sums. It may also be necessary to identify the risks carried by each party to determine whether some costs are to be excluded. For example, does the employer pay all costs if bad weather delays work? In part this can be achieved by a close definition of which costs will be reimbursed and which will not. Examples of possible wording to achieve this are given in the IChemE 'Green Book' conditions (see Section 4.5), and in the Schedule of Cost Components included in the ICE 'Engineering and Construction Contract' (see Section 4.2(f)). Care must be taken to ensure the wording adopted is clear. For complex works it may be necessary to carry out a risk assessment to identify potential problems and allocate the risks to either party.

(e) Target contracts

These are usually cost reimbursement contracts as (d) above, but with an estimated target cost set for the works cost, and a fixed or percentage fee for the contractor's head office overheads and profit. If the contractor's expenditure exceeds the target he has to bear a proportion of the excess; if his expenditure is less than target he receives a proportion of the difference as a bonus. Thus there is a financial incentive to the contractor to be efficient and save costs.

But setting a fair target price can be difficult, and impossible if the amount of work to be done is unpredictable. If a target has to be revised, a dispute may arise between employer and contractor as to what the new target should be; this defeats the purpose of this type of contract. If the work is reasonably well defined, then a measurement contract is usually suitable. Consequently a target price contract is not appropriate for many jobs.

If the initial target is set as a result of competitive tendering, then the employer may feel some assurance that he is obtaining 'value for money'. But if the target is negotiated or later has to be varied, then the employer may feel that the contractor's knowledge of his intended methods and costs may enable him to add a margin in the target estimate to safeguard his position. This means that it is improbable that the target cost will ever be lower than the contractor's privately estimated bottom line price.

(f) Payment under design, build and operate contracts

Arrangements for payment under design, build and operate (DBO) contracts may be partly direct and partly by income derived from the project operation as described in Section 2.6(b).

3.2 Other payment provisions

(a) Price variation provisions

In times of inflation it may be advisable to include clauses within the contract which set out how the contractor is to be reimbursed his extra costs due to any inflation of prices after he has tendered. Without such a clause the contractor has to add a margin to his prices to cover expected inflation, so he runs a risk he might not have allowed enough while the employer runs the risk that he might have to pay more than the actual inflation increase.

Calculating extra costs due to inflation can be complicated and time consuming for the contractor and the employer's supervisory staff, so that often a formula using officially published indices of prices is used instead. The contract has to set out how such indices will be used (see Section 16.8).

(b) Payment terms

Most standard conditions of contract contain specific provisions for interim payments and require payment of interest to the contractor if the employer fails to pay on time. The timing of these interim payments as the work proceeds is of importance to both employer and contractor. A contractor has to lay out large sums of money to get work started, especially on overseas jobs; hence the earlier he can receive substantial payments the less he has to borrow from his reserves or the bank.

On the other hand the earlier the employer has to pay out money, the more interest he will have to pay on his borrowings to fund the project. Also, he cannot pay out too large a sum or he may not recover his payment should the contractor get into financial difficulty and be unable to complete the work. The same result can follow if he pays out too little, and forces a contractor who is not in a strong financial position into further financial difficulties (see also subsection (e) below).

In the UK the *Housing Grants, Construction and Regeneration Act 1996* set out terms which must be included in construction contracts (see Section 1.6). These include a requirement for a clear system of payments by instalments, notice of the payment due and a date for payment, and notice of any money withheld. Where a contract does not contain the requirements of the 1996 Act, *The Scheme for Construction Contracts Regulations 1998* applies, setting out details of the payment provisions required by the 1996 Act. Most engineering standard forms used by employers already contained terms largely complying with the 1996 Act.

(c) Bonus payments

An employer can include in a contract the payment of bonuses to the contractor for completion of the works, or stages of it, on or before the time or times stipulated in the contract. Provided the times set are reasonably achievable and do not encourage the contractor to skimp work, bonuses can be rewarding to both contractor and employer. Early completion can reduce borrowing costs for the employer since he can gain an income earlier from the project output. Early completion also suits the contractor, since his overheads extend over less time and his profit on the job thereby increases.

The problem with bonuses is that, if unforeseen conditions occur causing the contractor a delay not of his own making, there may be a dispute about how much extra time should be allowed to him. Bonuses should therefore be a reasonable amount; not so large that they put the contractor on a win or lose situation in respect of his whole profit.

(d) 'Ex-contractual' payments

These are payments made by an employer to a contractor which are not authorized by the contract. They are occasionally paid when a contractor has performed very much to the satisfaction of the employer but has shouldered some extra cost clearly not attributable to his own actions, such as exceptionally bad weather or some other misfortune outside his control. Only the employer can decide to make an ex-contractual payment, not the engineer or other person acting on his behalf; and the employer must himself have power to make the payment. Hence a private person or company may be able to make an ex-contractual payment; but a public authority will usually have no such power.

(e) Pre-payments

An employer will rarely make an unconditional pre-payment, that is, a down payment to a contractor at the start of the contract. He can, however, make early payment to the contractor for provision of offices, laboratory, and transport for the engineer's staff on site, etc. (see Section 15.10). These matters by no means cover the contractor's outgoings for his initial set-up, especially when the project is very large and overseas, so significant advance payments, secured by a repayment bond, are often allowed.

On the Mangla Dam project in Pakistan the contractor needed to purchase and bring a vast amount of constructional plant on site. To ease the financial burden on the contractor, the employer (in effect the government) agreed to purchase or pay for plant required by the contractor up to a value of 15 per cent of the contractor's tender price excluding contingencies. The employer recovered this expenditure by deducting it in instalments over the first 30 months' interim payments to the contractor under the contract. In this case the employer could obtain further security for his down payment by retaining ownership of the plant until he reimbursed his outlay on the plant.

3.3 Contractual risks arising during construction

Among the most common risks encountered during the construction of a project by a civil engineering contractor under a standard type of construction contract, are the following:

1. Design errors, quantification errors.
2. Design changes found necessary, or required by the employer.
3. Unforeseen physical conditions or artificial obstructions.
4. Unforeseen price rises in labour, materials or plant.
5. Theft or damage to the works, or materials and equipment on site.
6. Weather conditions, including floods or excessive hot weather.
7. Delay or inability to obtain materials or equipment required.
8. Inability to get the amount or quality of labour required, or labour strikes.
9. Errors in pricing by the contractor.

Most standard conditions of contract apportion the normal risks of construction to the party best able to control the risk. The apportionment will vary from form to form but many have been agreed within the industry as giving a reasonable balance between employer and contractor and it is generally unwise to upset this for normal types of civil engineering work.

Thus under the ICE conditions of contract using a bill of quantities, risks 1, 2 and 3 are carried by the employer. Design changes can cause much extra work, cost and delay to a contractor but may be forced on an employer by circumstances outside his control. To safeguard his position, an employer should not enter unsuitable contracts which do not give him power to adopt reasonable design changes at reasonable cost.

Risk 4 is usually carried by the contractor in times of low inflation.

Risk 5 is carried by the contractor who has to insure against it, although the employer may also insure against consequent damage to works he owns, and to any new works he takes over.

Risk 6, delay due to weather conditions, has traditionally been a contractor's risk and this has posed many problems for contractors because the effect of inclement weather (mostly wet weather in the UK) can vary according to the type of work undertaken. Any form of earth or road construction can be severely affected by wet weather, whereas much building work need not be so affected. The ICE standard conditions entitle a contractor to an extension of the contract period for 'exceptional adverse weather conditions' but do not authorize additional payment on account of it. The ICE 'Engineering and Construction Contract' referred to in Section 4.2(f) attempts to define 'exceptional weather conditions' as a basis for claim, and allows time and payment if these are exceeded.

Risk 7, delay in obtaining materials, is carried by the contractor in most cases, except where the employer stipulates in the contract that a specific supplier shall be used, when liability for delay may lie with the employer (see Section 14.2).

Risk 8, failure to get labour, is usually shouldered by the contractor, mainly because this lies within the ability of the contractor to control, and not the employer.

Usually any requirement that the contractor should shoulder all or most risks arises because the employer prefers to have a fixed financial commitment, or because he has only a limited allocation of funds which he has no authority to exceed. Some overseas governments will not authorize any expenditure above the tendered sum. This fixing of the price and placing all or most of the risks on the contractor can be expected to lead to generally high prices. Also if unexpected circumstances occur which the contractor has not allowed for, he may tend to adopt 'short-cut' methods which do not produce the most satisfactory work, or he may be forced to finish the work at a loss.

For complex projects, and perhaps where major cost reimbursement or target cost projects are envisaged, a formal risk assessment may be necessary in which a risk register is set up, defining how each risk is to be dealt with and which party is to carry the liability should the risk occur. In summary this may involve:

- identification of risks likely to arise by discussion between all interested parties involved;
- analysis of each risk as to likely frequency, severity of impact on cost and delay, both maximum and minimum;
- identification as to who is best able to manage the risk and/or who should carry the costs which may arise;
- definition of risks falling on the contractor so that he can include for them in his prices or insure against them.

The analysis of risks may be accompanied by a mathematical probability exercise to try to assess the most likely outcome for the employer's financial planning purposes. As a general principle, it is usually best not to pass to the contractor risks which are most difficult to assess as regards likelihood or cost, since a contractor may then need to increase his price substantially to protect his position, causing the employer to pay for a risk which may never arise.

3.4 Producing an initial cost-estimate of a project

At an early stage an employer will want to know the probable cost of his intended project. Usually no realistic figure is possible until a feasibility study of the project has been completed; before that only an 'order of magnitude' figure or 'budget estimate' can normally be quoted. Three main methods of producing this are as follows:

- by reference to the cost of similar projects;
- by sketch layout and component costing;
- by use of cost curves if available.

The first assumes a record is available of the cost of past projects undertaken by the employer's engineer, or perhaps costs taken from the technical press.

The reference costs need to be accompanied by data, such as project size, project components and distinctive features, dates of construction, and whether the price includes land, legal and engineering costs. Inflation factors may have to be applied to update the costs. By comparing the principal features of the proposed project with those for which past costs are available, a probable order of magnitude total cost may be derived.

The second method is the most reliable. Even before a feasibility study is undertaken it should be possible to sketch out the proposed project on some notional site if the actual site is not yet decided, so the layout and sizes of the various components required can be judged. The components can be roughly sized so that their possible cost can be estimated by comparison with price data held for similar structures. This procedure can also reveal costs for items which might otherwise have been missed.

The third method, using published cost curves, is not very reliable, because the data on which such curves are based is so frequently absent, and virtually every civil engineering project has some unique feature substantially affecting its cost. Hence costs expressed per unit of size or output can vary greatly. However, a cost curve can be used to show whether costs developed by the other methods seem realistic.

While any of the above methods will involve uncertainty, they can be useful in comparing different options for a scheme, provided uniform parameters are used. The final estimate of cost drawn up by the engineer should be based on current prices and include a substantial contingency sum. It need not include for possible future inflation of prices, because this is a matter for the employer's financial advisers to deal with, but the basis of the estimate should be clear. The possible range of the cost should be shown; but whether the employer chooses to quote the highest or lowest estimate is up to him. Many a major project providing a major benefit (including the Channel Tunnel) would probably not have been built if the initial estimate quoted for it by the employer had not erred on the optimistic side.

3.5 Estimating the cost of a project at design stage

As the design of a project is developed a more accurate estimate of cost is possible, based on cost parameters derived from analysis of recent priced contracts for similar work. The designs should show the layout and sizes of component works required. For each such component it should be possible to make an approximate estimate of the quantities of the key structural operations required, such as bulk excavation, main concrete in framework and floors, wall areas, and roof areas. Examination of recent priced contracts can then produce cost parameters that can be applied to the estimated quantities for the proposed structure.

For example, using a past priced contract, the total concrete costs (inclusive of formwork, reinforcement, finishes, joints, etc.) can be divided by the volume billed in framework and floors to give a parameter to apply to the proposed building. Similar all-in cost parameters can be produced for all excavation (based on the bulk excavation); for exterior walls and windows (based on area); roof (based on area), etc. Having produced a total cost for these principal items, all other incidental costs for a structure can be expressed as a percentage on.

Pipeline costs can be expressed as per 100 mm diameter of pipe per metre laid, divided into supply and laying. Overall unit prices can also be developed for checking purposes, such as the cost of a building per m^3 volume; or of a tank per 1000 m^3 storage capacity.

Before the cost parameters are derived from previous priced contracts the following procedures are necessary.

- Preferably at least three priced contracts should be analysed. If possible not all should be for the lowest tenders received.
- Preliminaries and overheads (see Section 15.10) should be expressed as a percentage addition to the total of measured work.
- If a tender is being analysed, general contingencies and dayworks should also be expressed as percentage additions, or shown separately. If a final account is being analysed, then all non-identifiable payments for extras, dayworks and claims should be included in the percentage on.
- Special costs for special circumstances should be separately noted, to decide whether they apply to the proposed project.
- Prices obtained should be brought up to date by applying a suitable inflation factor. In the UK published indices of price fluctuation in UK construction costs are available (the Baxter indices) and overall price movements for different types of construction are tabulated in the government's *Monthly Bulletin of Indices.* If these are not to hand, good indicators for updating costs are current dayworks rate for skilled tradesmen and current prices for C25 grade concrete and reinforcement, as compared with those in the priced contract being analysed.

The advantages of the method are that the costs are real (i.e. as tendered), oncosts are included, and the procedure facilitates checking of costs by different methods. The sum total cost derived needs checking to ensure it appears reasonable.

During the design stage it may be found that a previous estimate appears too low; but it is important not to take over-hasty action in reporting this to the employer. The problems causing the increase should be examined first to see if some savings are possible. If an estimate must be increased it is better to do this only once, because a series of increased estimates may cause an employer to lose confidence in any estimate presented to him.

As the design nears completion more accurate estimates can be produced using the quantities taken off to prepare tender documents. Such quantities can be priced from historical data derived from priced contracts as indicated above, or, if necessary, from various 'Price Books' published.

It is important that this estimate is produced before tenders are invited. This gives the employer an opportunity to decide whether the cost is acceptable and, if not, to make some deletion to reduce the cost before the contract goes out to tender. The estimate can also act as a guide when comparing tendered prices.

3.6 Project cost control

It is during the design stage that measures to keep the cost of a project within a budget figure are most effective. All possible savings in design need to be sought, not only because this is manifestly in the interests of the employer, but because there are sure to be some unforeseen extra costs that need to be offset by any savings that can be made. Alternative designs of layout or of parts of the works have often to be studied before the most economic solution is found; hence completion of all design before starting construction makes a major contribution to controlling project cost.

The most prolific causes of extra cost are:

- not completing the design of the works in all essentials before the contract for construction is let;
- not allowing adequate site investigations to take place;
- encountering unforeseen conditions;
- making changes to the works during construction.

The first two listed above can be avoided by taking the appropriate measures. The third, however, is not avoidable even if the site investigations have been as reasonably extensive as an experienced engineer would recommend. The last – changes during construction – can be minimized by ensuring designs are complete before construction commences, and that the employer takes time to assure himself that the works as designed are what he wants. But some changes are unavoidable if, during construction, the employer finds changed economic conditions, new requirements or more up-to-date plant, or new legislation forces him to make a change. The designer should keep aware of possible changes to the employer's needs and other technical developments, and not so design the works that possible additions or alterations are precluded or made unacceptably expensive.

If tenders are received which exceed the budget estimate by so large a sum that the employer cannot accept any tender, means of reducing the cost may have to be sought. Generally speaking, down-sizing a part or the whole of the works is usually not as successful in reducing costs as omitting a part of the works. Reducing the output of some works or the size of a structure by 25 per cent, for instance, seldom results in more than 10 per cent saving in cost, and can make restoration at a later date to the full output or size an expensive and uneconomic proposition. If the employer can find some part of the works which can be omitted, this is a more secure way of reducing the cost of a project, and it should be possible to negotiate such an omission with the preferred tenderer.

4

Contract conditions used for civil engineering work

4.1 Standard conditions of contract

Over a period of many years there have been a large number of standard forms of conditions of contract introduced. Sometimes these have been developed by particular industries or specialist suppliers, but conditions for more general use have been developed by the major engineering and building institutions, as well as by government and allied organizations. Use of these standard conditions is beneficial because they are familiar to contractors, give greater certainty in operation, and reduce the parties' exposure to risk. Such conditions are often produced by co-operation between contractors' and employers' organizations, with the advice of engineers and other professionals experienced in construction. The documents thus drawn up give a reasonable balance of risk between the parties. However, their clauses are often interdependent, hence any alteration of them must be done with care, and is generally inadvisable because it may introduce uncertainties of interpretation. The main standard conditions used for civil engineering projects are listed below, with an indication of their main provisions.

4.2 Contract conditions produced by the UK Institution of Civil Engineers

(a) ICE Conditions of Contract for Works of Civil Engineering Construction

These are generally known as the ICE conditions and have for many years been the most widely used conditions for UK civil engineering works. They have a long history of satisfactory usage and have been tested in the courts and in arbitration so that the parties to a contract can be confident as to the meaning and interpretation to be placed on these conditions.

The latest edition is the 7th, published in 1999 together with guidance notes, reprinted with amendments in 2003. This edition is known as the Measurement Version to distinguish it from other ICE types of contract based on this established standard.

The principal provisions of the Measurement Version are as follows:

- The contractor constructs the works according to the designs and details given in drawings and specifications provided by the employer.
- The contractor does not design any major permanent works, but may be required to design special items (such as bearing piles whose choice may depend on the equipment he owns) and building services systems, etc.
- An independent engineer, designated 'the Engineer' is appointed by the employer to supervise construction, ensure compliance with the contract, authorize variations, and decide payments due; but his decisions can be taken by the employer or contractor to conciliation procedures, adjudication and/or arbitration.
- The contractor can claim extra payment and/or extension of time for overcoming unforeseen physical conditions, other than weather, which 'could not ... reasonably have been foreseen by an experienced contractor' (Clause 12) and for other delays for which the employer is responsible.
- Payment is normally made by re-measurement of work done at rates tendered against items listed in bills of quantities, which can also include lump sums.

A particular advantage of the ICE conditions is that interpretation of the provisions of the contract lies in the hands of an independent Engineer, who is not a party to the contract, but is required to 'act impartially within the terms of the contract having regard to all the circumstances' (Clause 2(7)). This gives assurance to both employer and contractor that their interests and obligations under the contract will be fairly dealt with. Also the contractor is paid for overcoming difficulties he could not reasonably have foreseen. Both these matters reduce the contractor's risks, making it possible for him to bid his lowest economic price. This benefits the employer, since the initial price is low and he does not pay out to cover risks which may not occur.

The ICE conditions contain many other provisions that have stood the test of time. These include requirements for early notice of potential delays and problems such as adverse ground conditions and provisions for submission and assessment of claims and valuation of variations. Properly drawn up and administered, a contract under these conditions appears fair to both parties, and the percentage of contracts ending in a dispute which goes to arbitration is very small.

(b) ICE Conditions for Ground Investigation

These conditions are based on those described under (a) but allow for the investigative nature of the work and the need for reports and tests. A schedule of rates may be used instead of a bill of quantities (see Section 3.1(a)). The need for a maintenance period and for retention money is left to the drafter, and will depend on whether permanent works, such as measuring devices, are included. The existing (1983) edition is now out of date and a new version is being drafted for issue in 2003 with provisions for dealing with any contaminated land discovered.

(c) ICE Minor Works Conditions

These are a shorter and re-written form of the ICE conditions (a) above for use on works which are fully defined at the tender stage and are generally of low value or short duration. The conditions can be successfully used for larger works, but the standard ICE conditions cover many more of the potential problems that can occur on more complex or longer-term projects. Payment arrangements are left open to be chosen prior to tendering, but are suitable for a single lump sum bid or priced bill of quantities. The 3rd edition of these conditions was published in 2001.

(d) ICE Design and Construct Conditions

These conditions were newly produced in 1992 with a 2nd edition in 2001. Known as 'the design and construct (D&C) conditions' they follow much of the wording of the Measurement Version but differ significantly from many of the principles of that version. Some of the principal differences are the following:

- The employer sets out his required standards and performance objectives for both design and construction in a document entitled 'the Employer's Requirements'.
- The Contractor develops these requirements and designs and constructs the Works in accordance with them.
- The Contractor is responsible for all design matters except any specifically identified in the Contract to be done by others.
- An 'Employer's Representative' is appointed who supervises the design and construction on behalf of the employer to ensure compliance with the Requirements and that the purpose of the works will be met. He has many duties similar to those of the Engineer under the ICE Measurement Version and is required to behave impartially in regard to certain decisions (Clause 2(6)).
- The Employer's Representative can issue instructions to vary the Requirements in reply to which the contractor must submit a quotation for any extra cost or delay in complying with these.
- Payment is normally on the basis of a Lump Sum payable in stages, although other means of valuation can be included. However, care is needed if work is re-measured against billed rates, since the contractor could then choose to adopt forms of design that suit the more profitable bill rates he has quoted.

D&C contracts are often used when the employer's main interest is to have some works built as soon as possible, and he need not, or does not wish to be concerned with the details of the design (see Section 1.3). The contractor can therefore start construction as soon as he has enough design ready. But where a project offers a wide range of design options, a design and construct contract may not offer an employer the best service because the options chosen by the contractor may tend to be those which suit his plant and workforce best, rather than the interests of the employer. However, if the 'Employer's Requirements' are sufficiently extensive and carefully specified, they can go a long way to ensuring coverage of all the employer's needs. It should be the aim of the parties prior to award of contract to arrive at an agreed scheme and specification for the works. Since this form of contract requires extensive input by tenderers their number should be limited to three or four only.

(e) ICE: Term Version

Term contracts have been in use for many years typically to cover repair and maintenance of facilities such as road surfaces or flood defences. This new form, based on the ICE 7th edition and issued in 2002, sets out a background contract which stays in place for a prescribed term of years and under which the Engineer can instruct packages of works to be undertaken as necessary. Works are ordered through a Works Order which defines what is required and supplies any drawings or specifications not already in the contract. Payment is made by measurement from a schedule of rates in the contract or other agreed means.

The administration and supervision requirements usually follow those of the Measurement Version and will thus be familiar to most engineers. This form can provide a welcome flexibility for employers in procuring irregular items of work or carrying out services which can be called up as and when needed and at short notice. The form may be suitable for some types of framework arrangement (see Section 1.12).

(f) ICE Engineering and Construction Contract

This contract was developed from 'the New Engineering Contract' (NEC) which was introduced in 1991 and substantially revised in 1993. The NEC is 'a family of contracts' comprising versions for construction, sub-contracted works, provision of professional services, and appointment of an adjudicator. The main construction contract was developed and renamed the **Engineering and Construction Contract (ECC)** which went into a second edition in 1995.

The ECC is formed from 'core clauses' which set out the general terms of the contract, 'main option clauses' which define valuation and payment methods (one of which must be chosen), and 'secondary options clauses' for such as bonus, delay damages, and price adjustment. A short form for minor works and a short sub-contract form are also available.

The contract requirements are defined in separate sets of data – Works Information and Site Information supplied by the Employer, and Contract Data which set out various pieces of information depending on which options have been chosen.

A *project manager* appointed by the *employer* administers the contract on behalf of the employer, assisted by a *supervisor* on site. A separate *adjudicator* is appointed to whom the contractor (but not the employer) can take disputes with the project manager or the supervisor for adjudication. But if the employer or the contractor disagrees with the adjudicator's decision either can have the dispute referred to any final tribunal set out in the contract.

The contract attempts to overcome some old problems by several new approaches, but the latter may present some new difficulties. A list of eighteen *Compensation Events* is prescribed, each of which entitles the contractor to claim extra payment and delay. They include the usual matters of claim such as variation of work, unforeseen conditions, etc. but add unusual weather. The latter is defined as – Weather recorded 'within a calendar month ... at the place stated in the Contract Data ... which by comparison with the weather data, is shown to occur on average less frequently than once in ten years.' The weather data is that supplied by the employer in the Contract Data, and a 'weather measurement', could, for instance be rainfall. This definition could give rise to problems of interpretation and may lead to claims even when the weather causes no delay.

Another provision is that the contractor's claims when he experiences a compensation event take the form of *quotations* which the project manager can accept, return for revision, or reject by advising he will make his own assessment in lieu. The problems with this approach are discussed in Section 17.12 below. Strict time limits of 2 weeks apply to stages of action and response by both contractor and project manager in respect of such quotations and other submissions. These times are tight and may cause difficulties; failure of the project manager to reply within a specified time limit being itself a compensation event!

The stated intent of the drafters of the ECC contract is to stimulate good management. This seems to be achieved by requirements for meetings in a variety of situations, so as to seek advantageous solutions to potential problems, and the tight timetables for responses between the parties.

(g) Partnering Addendum

This addendum has been issued by the ICE in 2003 to provide for the partners setting down their objectives and any risk sharing provisions formally. The addendum acts as an addition to individual contracts, which may be of any type, and allows for revision as partners leave or are added to a project.

4.3 Conditions published by the International Federation of Consulting Engineers (FIDIC)

FIDIC 'Red Book' Conditions, 4th Edition

The ***FIDIC 'Red Book' Conditions, 4th Edition*** are intended to apply to civil engineering work worldwide. They take the same form as the ICE conditions, but with some variations and simplifications to allow for work outside the UK. Additions can be made to cover local needs and different procedures for payment including payment in different currencies. The 4th edition – substantially revising the 3rd – incorporated changes resulting from consultations within the international construction industry and with major international lending agencies. These conditions were accepted by the major lending agencies who recommend or require their use together with additional clauses and amendments proposed by the agencies.

An important requirement in FIDIC4 is that the engineer is specifically required to consult with both the employer and the contractor before making a decision on a contractor's claim for additional payment or extension of the contract period. Another provision of importance is contained in Clause 52(3) which allows for adjustments to payment with respect to the contractor's overheads if the value of extra works ordered exceeds 15 per cent of the tendered sum excluding dayworks and provisional items.

1999 New forms

In 1999 FIDIC published four new forms. The first is a ***Contract for Construction*** to replace the Red Book. Much of the text and the concepts have remained but the whole is re-organized into what was considered a more logical sequence of clauses. The role of the independent engineer is retained who again has to consult with both parties to try to reach agreement on claims and the like, but if this is not possible, to 'make a fair determination in accordance with the Contract, taking due regard of all relevant circumstances'. The engineer's duty to make final decisions on disputed matters is replaced, however, by use of a Dispute Adjudication Board (DAB) selected by the parties.

A second form is for ***Plant and Design-Build*** which allows for the contractor to undertake design in accordance with the Employer's requirements set out in the contract. As the title suggests this is intended both for plant supply and installation and for use where much of the civil works may also be designed by the contractor. This contract again uses an independent engineer to monitor design and construction against the requirements and has the DAB to decide disputes.

Additionally FIDIC produced in 1999 a radically different form for engineer, procure and construct (***EPC***)/***Turnkey Projects***. Under this form the contractor takes over full responsibility for design and construction including any requirements of the employer, and undertakes to produce works which achieve the desired result. There is no independent engineer but an employer's representative who carries out various administrative and payment functions on behalf of the employer with disputes again referred to a DAB.

FIDIC have also produced a short form of contract for short-term projects of a fairly simple nature handled directly by the employer's staff.

4.4 Other conditions for civil engineering or building work

GC/Works/1 – General Conditions of Government Contracts for Building and Civil Engineering Works, Edition 3 (1991)

This edition is used mainly by UK government departments. They are, in consequence, widely used and are available in a number of different forms, for example, for payment by priced bills of quantities, lump sum, schedule of rates, or for design and construct, or supply only contracts. The contract is administered by a project manager or supervising officer who may be given powers similar to those of the engineer under the ICE conditions, but this depends on the policy of the government department concerned and type of work undertaken.

The employer (i.e. government department) takes on some powers exercised by the engineer under ICE conditions, including granting extension of time and deciding some payments to the contractor. Different departments may adopt different approaches in using the conditions, and new methods of contract administration have been tried out from time to time. Earlier editions of these conditions were felt to leave too much of the risk of construction with the contractor; for example by allowing neither extra time nor money in the event of bad weather. The 3rd edition of GC/Works/1 published in 1998 shows a more balanced approach but still does not require the project manager to act fairly.

Joint Contracts Tribunal Conditions

These conditions are not intended or used for civil engineering work but are the most widely used conditions adopted in the building industry; they are described here to show the building industry's different approach. Buildings will, of course, include many significant elements of civil construction, such as deep foundations or reinforced concrete structures such as a multi-storey car park. The 'Joint Contracts Tribunal (JCT)' which produces these conditions comprises representatives of the RIBA, RICS, ACE, various employers and building contractors and specialist contractors' organizations and representatives of local authorities. A range of standard forms of conditions provides for different types of employer and for payment by lump sum or quantities. Usually an architect or contract administrator supervises construction and issues certificates for payment, but a civil engineer may carry out these duties for structural works. Quantity surveyors, advisory to the architect, draw up bills of quantities and produce valuations and estimates.

Unlike civil engineering work, items in the bills contain descriptions of what is required in addition to any specification included in the contract documents, and the work has to be carried out in accordance with the bills and the drawings. Much of the work is carried out by sub-contractors appointed by the main contractor or sub-contractors nominated by the employer through the architect. The need for nomination arises so that the architect can obtain exactly the finishes, etc. he wishes to suit his designs. This tends to result in an increased possibility of disputes arising. A Clerk of Works may be appointed to supervise work on site for the employer but with very limited powers under the contract. It is thus possible for there to be three separate appointments – architect, quantity surveyor and clerk of works – taking part in supervision and this splitting of responsibilities and duties can lead to problems.

A Management form of JCT Contract was introduced in 1987 under which the onus for carrying out the work is placed upon a management contractor: that is, a firm of builders or civil engineers whose primary input is to manage and co-ordinate the inputs of sub-contractors (see Sections 1.10 and 2.5(b)).

4.5 Conditions mainly for plant and equipment supply

I Mech E Model Form A

This form, together with modifications that can be adopted (such as 'Form G' and a combined version called 'G90') is intended for contractor design, manufacture, supply, and installation of mechanical, electrical and instrumentation plant of all sorts. The form is still in use especially in its modified G90 form in the water industry despite the introduction of MF1 (see below). The contract allows for definition of what is required in outline and by specification, the contractor being responsible for the design and manufacture or procurement. The total plant required for a project is often procured by issue of contracts covering separate specialities, such as pumps and motors, switchgear, or instrumentation selected to suit the capabilities of tenderers. Provision for any associated civil works included in the contract is elementary: if they are required it is best they should be included as a fully designed package that can be sub-let. Payment terms are usually lump sum, but interim payments and some items of re-measure can be included. The terms provide for restricted liability of the contractor for defects other than during the first year of maintenance.

I Mech E/ IEE; MF/1

This is for similar purposes as Model Form A but the terms have been modernized and improved, with revised and extended liability for defects and provision for performance tests. The first edition was published in 1988 but further revisions with amendments have been issued, Revision 4 being issued in 2000. Among other matters, payment terms have to be decided and details added.

FIDIC 2nd and 3rd Editions: 'Yellow Book'

These apply to mechanical, electrical, instrument and similar work, the provisions being similar to I Mech E Form A conditions referred to above. The 3rd edition is a substantially altered and improved version of the 2nd edition. Again it is intended for work worldwide so it allows for additions to cover local needs. Although the new FIDIC form for plant is available this form is still in use.

I Chem E 'Red Book' Conditions

These conditions are primarily intended for process plant paid for on a lump sum basis, with interim progress payments as agreed. The contractor carries the main responsibility for design but must comply with any requirements set out in the contract. He arranges all procurement including any civil, mechanical, electrical and instrument work, etc. and installs, sets to work, commissions and tests all plant. Performance tests are required to prove that the effectiveness of the plant is as specified in the contract. Provisions for dealing with claims and variations are not extensive; the expectation being that, prior to the award of contract, the parties will have agreed in detail the specific items of plant to be provided, so that little change is needed. The project manager is not fully independent, though required to be impartial in some actions; the purchaser is bound by the decisions made by his project manager and cannot dispute them in arbitration. Provision is made for an independent expert to be called in to decide some technical or valuation matters. Since there is little allowance for unforeseen constructional problems, the conditions may not be suitable for major associated civil engineering works.

I Chem E 'Green Book' Conditions

These conditions cover a cost-reimbursable contract for the provision of process plant. They can be used when the process or works have not been fully defined, so both purchaser and contractor may have design inputs; but the contractor is responsible for proper construction. The contract can be on the basis of cost plus a percentage fee (rarely used) or cost plus a fixed fee, or a target cost. Payment terms have to define which costs are reimbursable and which are covered by any lump sum payments. Payment is made on the basis of the estimated expenditure for the forthcoming month, subject to adjustment of the preceding month's payment according to the actual expenditure incurred. This amounts to payment in advance. A project manager administers the contract but he does not have a fully independent status, although he is required to act impartially in some matters. The purchaser is responsible for the actions of the project manager and cannot dispute his decisions. These I Chem E conditions have been used occasionally for certain civil engineering works, primarily because they form a framework on which to base a cost reimbursement contract.

4.6 Other associated conditions

ACE Forms of Agreement

The Association of Consulting Engineers produces a series of forms of agreement suitable for the employment of a consultant or other person to carry out studies, design, construction supervision or management of various types. These agreements are compatible with the standard ICE forms of contract but are also intended for use alongside any of the major forms of construction contract setting out the obligations and liabilities of the consultant and defining his duties. Other forms of agreement are also available for project management, planning supervisor (CDM Regulations), designer to a D&C contractor and for sub-consultants.

CECA Sub-contract forms

The Civil Engineering Contractors Association produces a standard form of sub-contract for use with the standard ICE forms. Several versions are available to use with particular main contract conditions, such as ICE 7th edition and D&C. These sub-contract forms are generally known in the industry as the 'Blue Form' and care must be taken to identify which blue form applies to any specific sub-contract. Forms are also available for use with GCWorks 1.

The sub-contractor is deemed to have full knowledge of the main contract and many provisions such as for payment, claims and variations are intended to reflect the terms of the main contract. Mechanisms for dealing with disputes also reflect the standard ICE conditions and provision is made for these to be dealt with jointly where the matter in dispute affects both contracts.

5

Preparing contract documents

5.1 Initial decisions

Prior to commencing the preparation of documents for tendering and letting of a contract for construction, a number of decisions will have been made and actions taken.

- Feasibility studies and options for the project will have been concluded.
- Site investigations should have revealed ground conditions.
- Decisions will have been taken on the way design is to be carried out.
- A Planning Supervisor will have been appointed – for work in the UK – as described in Section 10.2.
- Decisions will have been taken on the breakdown of construction into different contract packages to suit capabilities of potential tenderers for civil works, plant suppliers, etc.
- Environmental studies will have been undertaken and requirements assessed.
- Planning submissions will have been made and other approvals sought.
- Initial project programmes will have been produced to indicate the sequences of construction.
- Financial planning will have been completed to ensure funds will be available.

From these considerations it should be possible to decide on the content of each contract as to the scope of works, whether or not the contractor is to carry out any design or whether designs will be provided complete for construction. These together with any financial requirements should enable a choice to be made of the conditions of contract likely to be most suitable for the works (see Chapter 4). One of the main aims in these decisions will be to try to ensure that contracts are packaged to suit the capabilities of the tenderers likely to be invited.

5.2 Roles of the key participants in a construction contract

A construction contract is made between two parties only – 'the Employer' and 'the Contractor'. Their roles are defined in the contract. However, because there is a need for day-to-day supervision of civil engineering construction, the two parties may agree that a third person should carry out such duties. This third person can have varying powers under the contract and this is reflected in his designation. He can be designated 'the Engineer' under the contract; or he may be designated 'the Project Manager' or 'Employer's Representative' in both cases occupying a distinctly different position from 'the Engineer'. The roles of these participants are described briefly below; the use of a capital letter in their designation being discontinued except where necessary for clarity.

The **employer**, referred to as 'the purchaser' in some conditions of contract, initiates the process of acquiring the works. He sets down what he requires and specifies this in the tender documents, which he issues to firms of contractors to seek their offers to carry out the works. His obligations include ensuring that the works are legally acceptable and practical, and that the site for them is freely available. He may also need to arrange that associated needs, such as the supply of power, drainage and the like which he is providing, are available. Having set up these basic elements he must, above all, ensure that he can meet his obligation to pay the contractor in accordance with the contract. If any dispute remains unresolved under the contract, the employer must decide what action to take; either to negotiate some settlement or, perhaps, take the dispute to adjudication, arbitration or the courts.

The **contractor** takes on the obligation to construct the works. In his offer to the employer he puts himself forward as being able to build the works to the requirements set out in the tender documents. In order to do this he will have studied the documents and any geotechnical or other information provided or otherwise available, visited the site and checked the availability of such labour, plant and materials as may be needed. Once his offer is accepted and the contract is formed the contractor takes on the obligation of doing all and anything needed to complete the works in accordance with the contract, regardless of difficulties he may encounter. He is responsible for all work done by his sub-contractors and suppliers, and any design work the contract requires him to undertake.

The **engineer** designated in the traditional form of contract under the ICE or FIDIC conditions described in Sections 4.2(a) and 4.3, has a role independent of the employer and the contractor. He is not a party to the contract; but he is named in it with duties determined by the parties. Although he is appointed (and paid) by the employer, he has to supervise the construction of the works as an independent person, making sure they accord with the specified requirements. He also acts as an independent valuer of what should be paid to the contractor, and as a decider of issues arising in the course of construction. The engineer will normally be an experienced and qualified professional whose knowledge and standing should be sufficient to assure both employer and contractor that the decisions he makes are likely to be satisfactory, and given independently and impartially.

In the most widely used conditions of contract, decisions made by the engineer can be accepted by the parties to the contract; but, if either party should so choose, the engineer's decisions can be challenged and if need be taken to external decision. This ability to challenge the engineer's rulings can be seen as supporting the effectiveness of his role (see Sections 8.2, 17.15 and 17.16).

Given efficient contract documents and completion of the designs before construction starts, the appointment of an independent engineer to administer the contract encourages contractors to submit their keenest prices. Many contractors will seek out the reputation of such an engineer for his experience and ability to apply fair dealing, and will adjust their prices accordingly. This benefits both the employer and contractor since it gives assurance that their interests will be protected. It also facilitates the resolution of any constructional problems that arise, so that disputes arising over contractors' claims are rare. Few civil engineering contracts handled in this manner need settlement by resort to arbitration or a court of law.

A **project manager** holds a different appointment from the independent engineer described above. His appointment is designated under the relatively recent ICE's ECC form of contract described in Section 4.2(f). He may carry out many similar duties as the engineer under the traditional form of contract, but he is not fully independent. The specific content of the contract will define the limit of his powers to act independently. Decisions made by the project manager on matters that are subject to assent by the employer will commit the employer, who will not be able to dispute them. From the contractor's point of view, the project manager's decisions will be regarded as the employer's; so he may feel it necessary to increase his prices to cover the risk the employer might tend to interpret the contract in his own favour. If the contractor is to offer his lowest prices he has to be assured the terms of the contract will be interpreted impartially; for this reason an adjudicator is appointed to whom the contractor can take his disputes with the project manager.

The **supervisor** under the ECC has a role which is mostly restricted to watching over construction and attending tests, etc. although he has some powers to issue instructions and for correction of defects.

The **resident engineer** is the engineer's representative on site under the ICE conditions or an assistant of the engineer under FIDIC. He may be delegated some of the engineer's powers depending on his experience and the type of work as well as the remoteness of the site from the engineer's office. His main role, however, is to ensure the works are carried out as required by the contract.

5.3 The contract documents

A contract is an agreement between two parties which they intend to be legally binding with respect to the obligations of each party to the other and their liabilities. The contract thus binds the contractor to construct the works as defined, and the employer to pay for them in the manner and timing set out. As civil engineering works are often complex, involving the contractor in many hundreds of different operations using many different materials and manufactured items, including employment of a wide variety of specialists, the documents defining the contract are complex and comprehensive. The task of preparing them for tendering therefore warrants close attention to detail and uniformity of approach, so as to achieve a coherent set of documents which forms an unambiguous and manageable contract. A typical set of documents prepared for tendering will include the following.

Instructions to tenderers

These tell the contractor where and when he must deliver his tender and what matters he must fill in to provide information on guarantees, bond, proposed methods for construction, etc. The instructions may also inform him of items which will be supplied by the employer, and sources of materials he should use (e.g. source of filling for earthworks construction, etc.).

General and particular conditions of contract

The general conditions of contract may comprise any of the 'standard' forms of contract mentioned in Chapter 4. The particular conditions adopted may contain amendments or additions that the employer wishes to make to the standard conditions. Usually the standard conditions (which are available in printed form) are not reproduced in the tender documents but they will be named by specific reference and a schedule will show what changes have been made to them.

The specification

This describes in words the works required, the quality of materials and workmanship to be used, and methods of testing to be adopted to ensure compliance. The specification usually starts with a description of the works to be constructed, followed by all relevant data concerning the site, access, past records of weather, etc. and availability of various services such as water supply, electric power, etc. Further details are given in Section 5.5.

Bill of quantities or schedule of prices

These form an itemized list covering the works to be constructed, against each item of which the tenderer has to quote a price. A bill of quantities shows the number or quantity of each item and its unit of measure, the rate per unit of quantity quoted by the tenderer, and the consequent total price for that item. This permits re-measure according to the actual quantity done under each item. Some bills contain many hundreds of items, classified by trade or according to a standard method of measurement; other bills contain a less number of items (see Chapter 15). A schedule of prices may comprise a series of lump sums or it may call for rates only, but can list provisional quantities which are estimated, that is, uncertain. They would be used, for instance, for a contract for sinking boreholes, items being provided for boring in stages of depth, the total depth to which any hole has to be sunk not being known in advance.

Tender and appendices

The tender sets out the formal wording which comprises the tenderer's offer to undertake the contract, the tenderer having to enter the sum price he offers. The appendices to tender will contain other matters defining the contract terms and which the tenderer confirms he accepts in making his offer, such as time for completion of the works, damages for failure to complete on time, minimum amount of insurances, completion of bond, etc. There may be other matters concerning the basis of his offer he is required to supply, such as currency exchange rates (for international contracts) or sources of materials.

The contract drawings

These should provide as complete a picture as possible of all the works to be built. The more complete the contract drawings are, the more accurately the contractor can price the work, and the less likelihood there is that variations and extra payments will be necessary. However, it is not necessary at tender stage to provide every detailed drawing that will ultimately be required (such as all concrete reinforcement drawings) so long as the contract drawings provided to tenderers show quite clearly what is required.

On small jobs all the foregoing documents may be combined in one volume; but on most jobs at least two and sometimes three or more volumes will be necessary. A tenderer is usually sent a second copy of the instructions to tenderers, bill of quantities, tender and appendices, so that he can keep one copy of what he has bid.

5.4 Bond, insurance, etc.

When preparing contract documents a number of matters of contractual importance must be considered. These will usually be dependant on the employer's preferences or any regulations under which he must operate, the type of work involved and the financial liabilities arising out of the work. Some of these matters are considered below. For the standard ICE conditions entries will be required in the appendix to the Form of Tender but for the ECC conditions it will be necessary to decide which options are to be included as part of the conditions of contract as well as making appropriate entries in the Contract Data.

The **Defects Correction Period** must be stated. This is the period during which the contractor must repair any defects in the works resulting from his workmanship. The period is usually 52 weeks for major construction but may be more or less depending on whether there is running plant involved or how soon defects may become apparent.

A **Performance Bond** is usually required where the employer feels he needs financial protection against a contractor failing to complete the works either due to lack of resources or financial instability. The size of the bond should cover having to re-tender and any extra costs to complete. The value should thus be chosen to suit the work but should not generally exceed 10 per cent of the value of works. Bonds are, however, relatively expensive to supply and thus put up prices and so may be considered unnecessary if the contractor is substantial, or part of a large group who can supply a parent company guarantee.

Insurance against third party risks needs to have a value set with reference to the likely risks of damage. In a rural area this may be small, but works near a major industrial complex may need a higher level of cover. Normally it is the contractor who provides insurance cover both for the works and third parties but sometimes employers prefer to provide these insurances to save costs. This can lead to problems when the cover provided does not fully represent the risks which arise.

The **Time for Completion** of the works should be set at a reasonable period given a proper level of resources. In many cases there will be a need for sections of the works to be defined and separate times for completion to be set for each of these. Sections may be needed if other contractors are to have access to the site (e.g. for plant installation) or if parts of the works are needed early.

Since the employer will be kept from using his works and may incur other costs if the contractor fails to complete on time it is usual to set **liquidated damages** to compensate him for his loss. These must be calculated as a genuine pre-estimate of the loss based on the value of the works output and other costs anticipated.

Most contracts provide that a portion of the money due to a contractor each month be retained by the employer. This **retention** has to be set and is usually a percentage (often 5 per cent) of the value of works certified up to a limit. The intention is to provide cover for defects and outstanding work but retention is unpopular with contractors who point out that bonds and money yet to be certified for payment provide funds should this be necessary.

5.5 Writing specifications

In writing specifications care must be exercised to ensure consistency of requirements throughout and conformity with what is written in other documents.

This consistency can be promoted if one person drafts all the documents or, if parts are written by others, one person carefully reads through the whole finished set of documents. An inconsistency in the documents can give rise to a major dispute under the contract, having a serious effect on its financial outcome. Some guiding principles are as follows.

- The layout and grouping of subjects should be logical. These need planning out beforehand.
- Requirements for each subject should be stated clearly, in logical order, and checked to see all aspects are covered.
- Language and punctuation should be checked to see they cannot give rise to ambiguity.
- Legal terms and phrases should not be used.
- To define obligations the words 'shall' or 'must' (not 'should' or 'is to', etc.) should be used.
- Quality must be precisely defined, not described as 'best', etc.
- Brevity should be sought by keeping to essential matters.

It is not easy to achieve an error-free specification. It is of considerable assistance to copy **model clauses** that, by use and modification over many previous contracts, have proved satisfactory in their wording. Such model clauses can be held on computer files so they are easy to reproduce and modify to make relevant to the particular project in hand. Copying whole texts from a previous specification which can result in contradictory requirements should not be adopted. Entirely new material is quite difficult to write and will almost certainly require more than one attempt to get it satisfactory.

The specification has to tell the contractor precisely:

- the extent of the work to be carried out;
- the quality and type of materials and workmanship required;
- where necessary, the methods he is required to use, or may not use, to construct the works.

Under the first an informative description is given of what the contractor is to provide and all special factors, limitations, etc. applied. Under the second the detailed requirements are set out. The extent of detail adopted should relate to the quantity and importance of any particular type of work in relation to the works required. Thus the specification for concrete quality may be very extensive where much structural concrete is to be placed; but it may be quite short if concrete is only required as bedding or thrust blocks to a pipeline. A 'tailor-made' specification appropriate to the nature of the work in the contract should be the aim.

Repetition of requirements should be avoided. If requirements appear in two places, ambiguity or conflict can be caused by differences of wording. Also there is a danger that a late alteration alters one statement but fails to alter its repetition elsewhere.

The third of the items noted above needs careful consideration, as there may be dangers and liabilities involved in telling the contractor how to go about his work. Some methods may need to be specified, such as the requirements concerning the handling and placing of concrete, but these and similar matters should be specified under workmanship and materials clauses. Other directions on method should be given only if essential for the design. For instance, if it is necessary to under-pin or shore up an existing structure, the exact method used should not be specified for, if the contractor follows the method and damage ensues, the liability for damage may lie on the designer. Usually there is no need to specify a particular method, but there may be a need to rule out certain methods; for example, that the contractor is not to use explosives.

It is important to avoid vague phraseology such as requiring the contractor to provide 'matters, things and requisites of any kind', or 'materials of any sort or description', etc. Clause 8 of the ICE conditions is sufficient to put on the contractor the obligation to do everything necessary to complete the works – 'so far as the necessity for providing the same is specified or reasonably to be inferred from the contract.' Similarly the phrase 'excavation in all materials' is ineffectual. The drafter might think it covers any rock encountered but it does not if the geological data supplied with the contract or reasonably available to the contractor provides no evidence of the existence of rock. Definitions such as those used in the Civil Engineering Standard Method of Measurement (see Section 15.3) should be followed. If there is evidence that rock might be encountered, a definition of it is required as discussed in Section 15.7.

5.6 Co-ordinating contracts for construction

Plant supply contracts

Many civil engineering projects incorporate electrical or mechanical plant which has to be ordered before construction commences because of the time required to manufacture the plant. Details and dimensions of the plant will be required to permit design of structures to proceed. Thus the employer has to let contracts for the supply of such plant in time for delivery of the plant to occur when the construction needs to incorporate it. Figure 5.1 shows a plan drawn up to co-ordinate plant deliveries for construction of a water treatment works. A plan of this kind is needed for many types of project.

However, because the employer orders the plant he becomes liable for any delay caused to the civil contractor if he does not place an order in time, or the plant supplier defaults on his promised delivery period. The resulting delay claim from the civil contractor can be expensive, hence it is prudent for the employer to allow a 'safety margin' on plant delivery times. Thus the delivery times for plant quoted in the construction contract are put somewhat later than suppliers' quoted delivery times to give a margin for possible delay. Although this may result in plant being ready before the construction contractor needs it, this is the best policy to follow. It is often possible to persuade a supplier to hold plant in store until needed, and the extra charge he may make for this – or

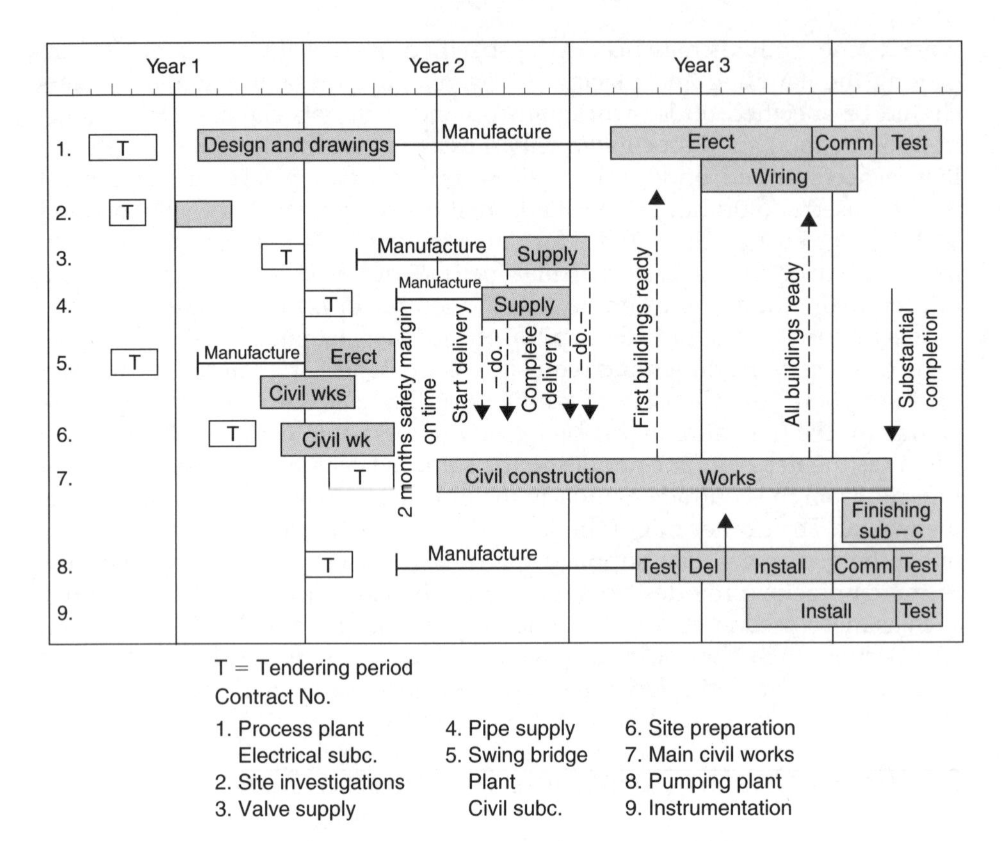

Fig. 5.1. Planning the co-ordination of contracts for a water supply project

the cost of getting the construction contractor to temporarily store some plant on site – will be less than the cost of delaying construction.

An alternative is for the employer to pass to the construction contractor the responsibility for arranging delivery of plant as he needs it. This was done on the Mangla dam project in Pakistan where eight major suppliers for hydro-electric plant, gates, valves and other large equipment were involved. The civil contractor was required to take over the plant supply contracts and arrange delivery to suit his construction programme, after such plant had been tested to the satisfaction of the employer. This kind of approach, however, is only practicable where the project is so large, as at Mangla, that construction takes several years to complete.

Site preparation contracts

An employer may also let a separate civil engineering contract for site-preparation, which covers building of access roads, bulk excavation, and providing electrical, water, and sewerage services to the site. Housing for site personnel may also be included. This contract can be let while detailed design of the project is still ongoing, and can therefore contribute to early project completion. Other advantages are that some excavation can be left open for the civil works tenderers to view, so minimizing the risk of claims for unforeseen ground conditions, and arrangements can be made for temporary storage of pipes and valves, etc. the employer orders which the main contractor has to incorporate in the works. However a site contract must be completed before the main civil engineering contract is let.

Co-ordination requirements

When separate plant supply contracts are let, the main civil contract must include all details of what the civil contractor must do in connection with such plant. Among the matters to be specified and allowed for in the bill of quantities are the following:

- Items requiring the civil contractor to take delivery of plant, offload, store, protect, and insure it.
- Items requiring the contractor to check deliveries of plant as invoiced by the supplier, inspecting items for any damage, and drawing the attention of the employer's engineer to any such, and to missing items.
- Where the plant supplier is to erect his plant, the main civil contract must state what services the civil contractor is to provide the supplier with, such as – access, scaffolding, lifting gear, power and lighting, water, use of the contractor's canteen and toilets, etc.
- Where pipes or other plant items have to be built in the civil works, the contract must make clear whether such items have to be built in 'as the work proceeds' or whether a hole can be left for a pipe to be 'built in after'. The contract drawings should show how the latter has to be done, and who is to be responsible for positioning any such item correctly.

The 'interface' between all separate contracts has to be carefully checked to ensure that all matters to be done by the plant supplier or civil contractor are properly covered and none missed out. The principal responsibility for this will lie upon the employer's engineer in charge of design and the drawing up of contracts. On a large scheme where several teams of engineers work on different parts of the design, checking that the interface between their separate parts match, is equally important.

5.7 The specification of general requirements

In Section 5.5 some problems of writing specifications have been mentioned. A specification usually comprises two distinct parts – Part 1: all the general requirements, and Part 2: the quality of workmanship and materials required. The general requirements can usually be classified into four categories:

- scope of work and reference standards;
- drawings and documents;
- site details and data;
- completion and testing.

Under the *first*, the specification should provide a brief but reasonably comprehensive description of the works to be built. The elements making up the whole project should be mentioned, together with their principal sizes or, where relevant, outputs. (This is of assistance to those who might wish to use the priced contract later for the purpose of analysing costs.) The services which the contractor is to provide may need description, particularly if he is to design any part of the works. The services which the employer and/or other contractors are to provide must be defined. Explanation should be given of the industry or national standards used on the project, and in what circumstances alternatives may be allowed.

The *second* section should include:

- a list of drawings provided by the employer to accompany the contract;
- requirements for any drawings and explanation of methods of construction the contractor is to produce, in order that sufficient information is provided for the employer to decide whether such work is as specified and conforms to all safety measures required;
- the timing of submission of the contractor's drawings and what time is allowed for the engineer to examine same and respond;
- other information required from the contractor such as – test results on materials and items of plant the contractor is to provide including manufacturers' drawings, maintenance and operation manuals;
- an example of the form in which claims for interim payment should be submitted.

The *third* section will contain much information about the site and relevant data, such as

- description of site and access, working areas;
- statutory requirements e.g. work in public roads, Health & Safety Act, Control of Pollution Act, etc.;
- water and power supplies available, sanitation, sewerage and solid waste disposal;
- contractor's offices;
- engineer's offices, attendance on engineer, vehicles for engineer, telephones;
- temporary fencing, watching;
- setting out data;
- geological and hydrological data.

The geological and hydrological data presented is of crucial contractual significance. The contractor has to base his prices on what is reasonably foreseeable;

hence he must seek all data which has some relevance to what might be foreseeable. The ICE conditions 7th edition contain Clause 11 which deems that the contractor has based his tender on his own inspection and examination of the site 'and on all information whether obtainable by him or made available by the Employer'. The employer should thus supply to tenderers any information he has which can be considered relevant to the works.

Choice of this information can present serious problems. There may be a large amount of such data, and it may be of variable reliability due to use of different methods of procurement, testing samples, etc. But to hold any data back on the basis of its doubtful validity would be dangerous; it could turn out to be highly relevant to troubles the contractor might encounter. Yet to comment on the reliability of data, would be equally dangerous. Clause 11 states that 'interpretation' of the data is the contractor's responsibility. Therefore the data included in, or supplied with, the tender documents must be chosen with care by the geotechnical engineer in charge of such matters and no interpretation or comment on such data should be given. However factual descriptions of the methods used for obtaining data should be given because this is relevant information which can indicate variations of data reliability. The dates when investigations took place and their exact locations are also essential information.

The *fourth* section defines any requirements or restrictions in respect of the programme for construction and the completion of the contract, including details of sections of the works required to be completed early. If any bonus is to be allowed for completion of all, or some part of the works by a given date, this should be defined. Details should be given of all other contractors who will have rights to enter the site, what work they will undertake and what facilities they will need. Lists of contractors supplying materials to be incorporated in the works need to be given, together with expected times of delivery. Tests stipulated before work can be accepted and should also be detailed.

5.8 The specification for workmanship and materials

Part 2 of the specification will cover workmanship and materials, and will often be lengthy, perhaps comprising a volume on its own for a complex project. It is usual to specify a material and its associated workmanship together in the same section. If workmanship is described separately from materials there is a risk that some workmanship requirement may be overlooked by tenderers.

The specification is normally divided into classes of work or trade. One method is to take trades in the order they are listed in the ICE standard method of billing quantities (CESMM). But CESMM lists 'miscellaneous metalwork' (Class N), and 'softwood components' (Class O), before 'piling', 'tunnelling' and 'engineering brickwork' (Classes P, T and U respectively) – which is not the order in which construction normally proceeds. An alternative is to list trades, both in the specification and bill of quantities, in the order in which they will be used. This is more logical, helps drafting and makes sure matters are not missed.

In drawing up the specification it is advisable to plan out beforehand subjects to be dealt with. An order such as the following might be adopted.

1. demolition, site clearance;
2. excavation;
3. piling;
4. concrete;
   (a) in situ;
   (b) reinforcement;
   (c) formwork;
   (d) pre-cast;
   (e) pre-stressed;
5. pipe-laying (might be put later);
6. steelwork (structural);
7. brickwork/blockwork/masonry;
8. roofing;
9. cladding (if special);
10. carpentry;
11. finishing trades (as necessary);
12. roads, site restoration, fencing.

It is not advisable to use three or more levels of decimal numbering of sections, such as '2.1.1', '2.1.2', etc. Only the section headings under each class of work need be numbered; sub-sections can be un-numbered and identified by a left hand heading, and paragraphs are not numbered. This permits insertion of late additions without disturbing any numbering.

When drafting the specification, care should be taken to ensure coverage of all types of work that appear on the drawings. In civil engineering contracts the specification sets out all quality requirements so these must be complete. The items in the bill of quantities only need sufficient description for the item to be identified for the purposes of payment. If an item in a bill of quantities appears with extra description which is not in the specification, a contractor might argue that the item with the added description requires additional payment. (See Section 4.4 where it is noted that under JCT conditions for building work the contrary practice applies.)

There have been differences of opinion among engineers as to the merits of 'method' as against 'performance' types of specification. A method specification for concrete quotes not only the materials and quantities of them to be used to make various grades of concrete but also the strengths and other physical characteristics to be achieved together with requirements for handling and placing. A 'performance' specification would stipulate only the strength and other physical characteristics to be achieved. This, it is said, leaves the contractor greater freedom to decide how he will achieve the performance criteria. However, opponents of performance specification point out that control by testing is only possible (in the case of concrete) 28 days after placing, and such tests may not provide sufficient proof that the structure will perform satisfactorily in the long term. If defects appear later, how is the contractor to be held responsible? Traditional 'method' specification is therefore the usual practice adopted, based on long-standing practices that have proved satisfactory over many years of experience.

The specification ought to be relevant to the work required. To include provisions that are irrelevant shows signs of a careless approach. When complex matters have to be specified it is best to avoid long and complicated sentences; short sentences are better. Occasional lists of requirements can aid clarity. The specification is a reference book which should be easy to consult. Inevitably large parts of it will cover such obvious requirements that they will not be read – for example, 'All formwork must be true to line and level'. Any unusual or special requirement should therefore not be tacked onto such standard material, or it will get missed. It should be put as a separate paragraph, even if it comprises only a couple of lines, so that it stands alone.

When writing specifications much use is made of past experience. Many engineers and consultancy firms will have model clauses available for specifying materials in common use. Both 'short version' and 'comprehensive' model specification clauses may need to be drawn up for a given material, so that the appropriate model clause can be used, according to the amount and importance of that material in a job. Considerable use will be made of national standards, standard sizes or qualities of manufactured goods. But it should be noted that, although British or other national standards are often quoted, this may not be sufficient definition because many such standards cover various grades and qualities, and precise references may be needed. Permitted alternative national standards may need to be quoted also.

Use can be made of a trade-named product to specify a material required but, wherever possible, it is better to avoid restricting the contractor to just one product by adding after the named product the words 'or similar approved'. The problems caused by nominating one supplier are dealt with in Sections 14.2 and 15.8, and it must be borne in mind that the practice could be contrary to EU competition rules.

Many sections of the civil engineering industry have their own approved technical specifications which are meant to act as standards for their particular type of work. The UK Department of Transport (Highways Agency) has published extensive clauses covering all manner of roadwork in their Highways Specification. The water companies in the UK have published a Civil Engineering Specification for the Water Industry. These documents can give an important lead to the specifier, but should not be slavishly copied, but checked, amended and extended so as to relate concisely to the needs of each particular job.

It is wise to enquire of the employer whether he wishes any particular specifications or standard requirements to be adopted. This is important when work is undertaken for governments or public utilities overseas. They often have their own printed specification, departures from which may not be permissible since they might not be noticed or understood by local tenderers. The sanction of the employer might be needed before any addition or amendment is made to such traditional specifications and, if allowed, may need to be put in a special section and carefully worded in simple English.

6

Tendering

6.1 Methods used for obtaining tenders

An employer usually seeks bids or 'tenders' from construction contractors on the basis of tender documents produced as described in Chapter 5. Tenders can be obtained in one of three ways – by 'open' tendering, selective tendering, or by negotiation.

Under **open tendering** the employer advertises his proposed project, and permits as many contractors as are interested to apply for tender documents. Sometimes he calls for a deposit from applicants, the deposit being returned 'on receipt of a bona fide tender'. However, this method can be said to be wasteful of contractors' resources since many may spend time preparing tenders to no effect. Also, knowing their chances of gaining the contract are small, contractors may not study the contract in detail to work out their minimum price, but simply quote a price that will be certain to bring them a profit if they win the contract.

Thus the employer may be offered only 'a lottery of prices' and not necessarily the lowest price for which his project could be constructed. If he chooses the lowest tender he runs the risk the tenderer has not studied the contract sufficiently to appraise the risks involved; or the tenderer might not have the technical or financial resources to undertake the work successfully. It is true that the employer can check the resources and experience of the lowest bidder and reject his tender if the enquiry proves unsatisfactory; but several bids may be below the estimated cost of the job and, if such tenderers appear satisfactory and their bids are not far apart in value, it is difficult for the employer to choose other than the lowest. The engineer advising the employer may think there is a risk that all such low bids could prove unsatisfactory, but he cannot advise the employer what other bid to accept because he has no certainty of information.

Under **selective tendering** the employer advertises his project and invites contractors to apply to be placed on a selected list of contractors who will be invited to bid for the project. Contractors applying are given a list of information they should supply about themselves in order to 'pre-qualify'. The advantage to the employer is that he can select only those contractors who have adequate experience, are financially sound, and have the resources and skills to do the work. Also, since only half a dozen or so contractors are selected, each contractor knows he has a reasonable chance of gaining the contract and therefore has an incentive to study the tender documents thoroughly and put forward his keenest price. However, since contractors have all pre-qualified it is difficult to reject the lowest bid, even if it appears dubiously low – unless that is due to some obvious mistake.

A problem with both open and selective tendering is that a contractor's circumstances can change after he has submitted his tender. He can make losses on other contracts which affect his financial stability; or may be so successful at tendering that he does not have enough skilled staff or men to deal with all the work he wins. Neither method of tendering nor any other means of procuring works can therefore guarantee avoidance of troubles.

Negotiated tenders are obtained by the employer inviting a contractor of his choice to submit prices for a project. Usually this is for specialized work or when particular equipment is needed as an extension of existing works, or for further work following a previous contract. Sometimes it can be used when there is a very tight deadline, or emergency works are necessary. A negotiated tender has a good chance of being satisfactory because, more often than not, it is based on previous satisfactory working together by the employer and the contractor.

When invited to tender the contractor submits his prices, and if there are any queries these are discussed and usually settled without difficulty. Thus mistakes in pricing can be reduced, so that both the engineer advising the employer and the contractor are confident that the job should be completed to budget if no unforeseen troubles arise. However, negotiated tenders for public works are rare because the standing rules of public authorities do not normally permit them. But a private employer or company not subject to restraints such as those mentioned in the next section can always negotiate a contract, and many do so, particularly for small jobs. Even when a negotiated tender is adopted it is usual to prepare full contract documents so that the contract is on a sound basis. Production of the documents also means they are available for open or selective tendering should a negotiated tender fail, or should the chosen contractor be unable to undertake the work.

6.2 Tendering requirements and EC rules

Civil engineering construction works and many other similar types of purchase form a large part of the annual expenditure of local and national government authorities and of the public services such as water, drainage, gas and electricity, etc. Consequently such authorities have long-standing rules concerning the procurement of tenders, designed to ensure tenders are obtained openly in a manner which gives best value for taxpayer's money. The rules may stipulate the number of contractors to be pre-qualified under selective tendering according to the size of contract to be let, and penalties to be imposed if bribery or collusion by tenderers is discovered. In the UK the government introduced 'compulsory competitive tendering' (CCT) into local government, national health and other public services, under which the employing authority's own direct labour force was required to bid in competition with outside contractors' offers in order to gain work. Initially the competition was on tendered price, but later was modified to be on 'Best Value', that is, on the quality of materials, workmanship, design, etc. of tender offers (see Section 1.15). The UK Local Government Task Force later published a guide for local authorities containing some 120 recommendations for procedures advisable when commissioning a construction project.[1]

Under some European Community (EC) Directives rules have been set for tendering procedures for construction work, and also for the supply of goods and services, which members of the Community are obliged to follow. An appendix to this chapter lists the UK Regulations implementing the EC Directives. The EC has also been considering how bids for contracts can be submitted by electronic communication means, that is, 'E-procurement'. An EC study group (IST 1999/20570) of contractors was set up to advise and report by 2002 on the framework needed for specifying the legal conditions for use of electronic communication.

The rules vary according to the expected value of the contract and the type of service required. Most supply contracts for public works come under EC rules if they exceed a threshold value of 400 000 ECU (£250 000 approximately); while construction contracts for the utilities (e.g. gas, water, drainage, electricity, etc.) come under control for contract values exceeding 5 million ECU (£3.2 million approximately). Limitations are applied to prevent splitting down work to avoid such rules. Some general features of the EC rules are as follows:

- *Procedure.* Open, restricted (i.e. selective), or negotiated tenders may be sought. Open tendering requires public advertisement inviting tenders for a contract. Restricted or selective tendering, requires public advertisement inviting tenderers to pre-qualify for some specified type of work, then inviting bids for work of that type from some or all of pre-qualified tenderers. For a negotiated tender at least three tenderers from those pre-qualified must be invited to bid – except in certain circumstances, such as for emergency work, for research or development, or for technical or artistic works available from only one source, or as an unforeseen addition to a current contract or a repetition of some current contract work.
- *Advertising.* Work coming under the rules must be advertised annually, or for each contract individually, to permit tenderers to register their interest.

[1] *Rethinking construction: an implementation guide for local authorities; implementation toolkit*, published by Local Government Task Force, 108–110 Judd Street, London WC1 – undated but issued mid 2001.

If pre-qualification is not used, a contract notice advertisement must be issued. Minimum times are set for lists to remain open. After the award of a contract, an award notice must be issued.
- *Contract award criteria.* The criteria which are to be applied in awarding a contract must be set out. The choice may be the lowest tender, or that which is economically the most advantageous. In the latter case the factors which will be taken into account when judging tenders must be stated and adhered to.
- *Publicity.* All required notices must be published in the Official Journal of the EC, using standard formats.
- *Standards.* Standards must not be set in specifications in a manner which restricts trade between Community members. European standards, or National standards which implement such standards, take preference over any others.

The EC is not the only body setting rules for tendering. The major international lending authorities such as the World Bank, Asian Development Fund, UNO, WHO, etc. also have rules to ensure that tenders are open to international contractors, or to contractors from countries funding a project, or to conform to some specific requirement. Individual countries often require that tendering procedures shall be so designed that either certain goods and services come only from inside the country, or maximum possible use is made of these. Limits may be set on the use of imported goods or services. On construction contracts tenderers may be required to work in conjunction with local contractors and the number of expatriates the contractor employs in the country may be restricted.

6.3 Procedures under selective tendering

If an employer is not subject to any of the restrictions outlined in the previous section, he may make a selected list of contractors from those who have served him satisfactorily in the past or those recommended to him. However, for public authorities in the EC, EC rules will apply for contracts above a certain value as described in the Section 6.2, and elsewhere in-country rules or rules set by an international funding agency may apply. For selective tendering, lists of potentially suitable tenderers who have pre-qualified under a previous selection process can be compiled. These can be standing lists reviewed perhaps annually, or lists compiled for specific types of work. Under EC regulations, selecting a list of pre-qualified tenderers on 'a framework basis' is also possible (see Section 1.12). The 'framework' either provides for such tenderers to bid for future contracts of a given kind, or permits direct selection of a tenderer on the basis of competitive rates already tendered.

Contractors wishing to be placed on a standing list may either answer the employer's advertisements or apply direct to the employer. This pre-qualification will usually seek to establish three categories of information about a contractor as follows.

- *The contractor's organization and resources.* Details of his ownership, details of staff available for the contract, and information concerning any special equipment or skills available for the particular type of work proposed.
- *Experience and performance record.* The experience the firm has of projects similar in type and size to the intended project, and what performance thereon was achieved. Some of this information may have to be obtained by asking the contractor to provide references from previous employers, the references being taken up. It is not always desirable to restrict the list to contractors who have done work of a similar kind and magnitude before, as this could unnecessarily restrict the choice of contractors and exclude competent contractors who have growing resources and skills.
- *Financial standing.* A contractor must be able to show he has sufficient funding to carry out the proposed contract without over-stretching his financial resources. The contractor may be asked for his turnover and recent financial history and data with respect to his current financial commitments. Some of this information may be available from annual financial reports or other sources; but it may be important to check that all relevant data has been supplied. An accountant may be employed to enquire into these matters.

In order to collect the necessary data in an organized manner it is preferable for standard forms to be issued to contractors, otherwise comparison and analysis may be hindered. A format suitable for international tendering is available from FIDIC, and guidance is also given by the World Bank in their Standard Bidding Documents. If the purpose of pre-qualification is the construction of a specific project, then applicants should be told the grounds on which their suitability will be assessed. Care is needed in defining these grounds. On the one hand the criteria applied need to be sufficient to keep those qualifying to a reasonable number; on the other hand, they should not be so tight as to exclude potentially suitable contractors who just fail to meet one of the criteria applied.

For works of a value up to about £1 million, a list of four to six pre-qualified tenderers would usually be regarded as sufficient; for larger value contracts it is seldom desirable or necessary to have more than eight pre-qualified contractors invited to bid. For design and build contracts a list of only three or four may be sufficient. Where a standing list is maintained, this can be divided into lists of contractors best suited to certain kinds and magnitudes of work, but contractors should be given reasonable opportunity to change their listing on supplying additional information. Once a selected list has been produced and approved by the employer, it is advisable that contractors on the list are approached individually shortly before sending out tender documents, asking them to confirm they are still interested in and capable of tendering. This is important if there is any substantial lapse of time between the pre-qualification of contractors and the sending out of tender documents. Contractors' commitments can change in a relatively short time. Those pre-qualifying should be informed of the expected timing of the issue of tender documents and start of construction, so that they can plan their tendering work and consider their response to other opportunities.

6.4 Requirements for fast completion

There is a widespread tendency for employers to want their civil engineering works designed and built in the shortest possible time. This pressure for speed often arises because of the time it nowadays may take for an employer to gain powers to construct his works. He may need to get planning consents, satisfy conservation and environmental interests, acquire land and wayleave rights, accommodate objectors and go through the lengthy process of a public inquiry. The funding of international projects may also take some years to arrange and negotiate. Commercial organizations tend to delay a project, then want it completed as fast as possible when market conditions are right. None of this can be avoided; but the pressure to undertake both design and construction in excessive haste needs to be resisted.

Starting construction before designs are complete or before the employer is sure what he wants, is a major cause of constructional problems, claims by contractors and of costs grossly exceeding original estimates. Hurriedly prepared documents, contract drawings incomplete before tendering, tender periods too short and an employer who wants changes after construction has started, can lead to a legion of unforeseen problems, forced changes of design, multiple claims from a contractor and a job not completed any earlier than it would have been had proper time been allowed to get everything right before calling for tenders.

The three outstanding requirements for fast completion of construction are:

- designs fully complete before tenders are called for;
- adequate tendering time for tenderers to prepare their bids;
- an employer who makes no substantial changes to his requirements after construction has started.

Given good designs based on adequate site investigations, drawings providing all the details a contractor needs, and no changes during construction, any competent and well-organized contractor can give fast construction. He can also give a good job. The quality of a job is all-important to a contractor's reputation. The cost of a job and how long it took may fade from an employer's memory; but, if the job is a poor one giving a series of after-troubles, it will be a continuing source of dissatisfaction to the employer, which he will not forget.

6.5 Issuing tender documents

Tender preparation and assessment times need to be adequate; they should be programmed into a realistic timetable which gives sufficient time for the engineer and the contractor to carry out their duties. A contractor faced with a set of contract documents has to absorb much information, get many quotations, and consider all options. For a small job even 4 weeks' tendering time may fall short of his needs; for major projects up to 3 months' tendering time may be needed to ensure that tenderers have time to consider all their strategies and put their best price forward.

Not less than two sets of documents should be sent to each of the contractors on a selected list and, if a substantial amount of specialist input is specified, further copies of the parts of the documents covering these specialist requirements ought to be supplied for the contractor to send to his specialist suppliers. Electronic supply of documents and drawings may help tenderers particularly if time is short or if suppliers are to be sought from around the world. Employers sometimes consider tenderers should pay for all sets of documents they receive; this may be prudent when open-tendering is adopted in order to prevent frivolous enquiries, the payment being returned to contractors who submit proper tenders. For selected tenderers, payment should be unnecessary except when a tenderer makes an unreasonable demand for extra copies. In whatever manner tender documents are sent to contractors, details of their despatch should be logged, and each tenderer should acknowledge receipt. All drawings and specifications should eventually be returned to the employer.

During the tendering period it may be necessary to issue amendments to tender documents. These may stem from errors and inconsistencies coming to light, queries raised by tenderers and changed requirements by the employer. Each amendment should be numbered, and a copy sent to every tenderer, with a request for him to acknowledge its receipt. If any query is raised by a tenderer (even by telephone) and the answer given to him provides him with additional or clarified information, this information must be confirmed in writing and the same information must be sent to all other tenderers; but the identity of the tenderer raising the query should not be disclosed.

Too many amendments should be avoided since they can cause disruption to tenderers and may lead to requests for extension of the tendering time. Minor errors found in the specification or drawings should not be circulated; they should be noted for correction at a later stage. If the employer requires some important change, careful consideration must be given as to whether tenderers should be advised of it, or whether the change should be held back to be dealt with when making the award of tender, or by issuing a variation order after the award of contract. Amendments should not be issued late in the tendering period. Requests received from contractors for extension of the tendering time can be avoided by giving adequate prior notice to selected contractors as to when tender documents will be issued, and giving sufficient tendering time. No extension of tendering time should be allowed if any tender has already been received, even though it remains unopened.

Sometimes **pre-tender meetings** are held which all tenderers are invited to attend; they are usually addressed by the employer who wishes to clarify some special aspect of the proposed project or give information concerning some important query raised by a tenderer. Preferably such meetings should be avoided because they can provide opportunities which undermine the independent nature of competitive bids. However, visits to inspect sites will need to be paid by tenderers. If such site visits are made in the company of the employer's engineer or one of his assistants, the engineer must be careful to provide only factual answers to queries raised. Should this provide a visiting tenderer with additional information this will need to be sent out to all tenderers. It is better if the tenderer visiting is accompanied, if need be, by a guide who is not directly connected with the contract, any queries being noted and dealt with formally after the visit.

6.6 Considering tenders

Opening tenders

Arrangements for return of tenders should be set out in the 'Instructions to Tenderers', giving both the place and latest time for receipt. Tenderers need to use secure means of delivery, and should receive a signed confirmation of delivery. It is usual to require tenders to be returned in sealed envelopes, marked only with the contract name and no means of identifying the name of the tenderer. Arrangements should be made to mark each tender envelope with the date and time of receipt, and for the safe storage of same until opening is authorized. Documents received after the closing time should be similarly marked and held unopened, until the employer decides whether they can be considered valid or not. Obviously common sense must be exercised; the employer will not wish to have a genuine bid invalidated by conveyance mishaps outside the control of a tenderer, such as a postal strike, or aircraft delayed. Once tenders are opened, no late delivery of a tender can be considered.

Tenders for large projects are sometimes opened at a public ceremony, the name and total tendered price of each tenderer being announced. This has the advantage that everything is 'above board' so that practices which could distort price competition are precluded. Also, contractors gain immediate knowledge as to how they stand with respect to getting the contract. In other cases, such as in local government, the practice is for tenders to be opened by a senior official in the presence of the chairman of the appropriate committee and others according to the standing rules of the authority. A record is usually made of the tendered prices as opened and signed by one or more of those present.

The tenders when opened are then usually passed to the employer's engineer for examination. The first step is to mark all documents with the name of the tenderer and list them. This list should be given an independent check so as to be certain that, if a tenderer says one of his documents has been missed, the employer's officials can show it was not received. Once the list has been compiled, any document not returned by a contractor that should have been returned, is noted.

Qualifications attached to tenders

Some tenderers may attach qualifications to their tender, usually set out in their covering letter. Qualifications which simply refer to some minor interpretation of a statement in the documents can usually be left for later agreement. But some qualifications may deal with a matter of considerable importance that changes part, or all, of the basis of contract as set out in the contract documents. Some employers have rules which require qualified offers of this kind to be rejected out of hand: if so, this should be made clear in the Instructions to Tenderers.

An offer which is qualified in some important respect may give a tenderer an unfair advantage over other tenderers. For instance, a qualification that tendered prices are subject to increase if rates of wages ruling at the time of tendering increase, would invalidate a tender if the contract documents contain no such provision. On the other hand, a tenderer may submit a more subtle qualification, such as 'Our prices are dependent upon being able to complete the contract within X months', where the months X are less than the period for completion stated in the contract documents. The problems this can lead to are discussed in Section 6.9.

A contractor can, of course, always submit an offer in accordance with the contract documents and add a second offer which proposes some reduction on the former offer if some qualifying condition is accepted.

Whether a qualified offer can be accepted or not depends upon the powers and restraints under which the employer operates. A private person or company, subject to no restraint, can accept any tender. But a public authority or utility will be bound by standing rules and perhaps EC or other rules also; while a project funded by one of the international funding agencies may come under rules that preclude any qualification being accepted which could be inferred as granting a favour to one tenderer and not to others.

In some cases specific alternatives are allowed in the contract documents. These most often refer to methods of constructing major temporary works, such as cofferdams for bridge foundations; river diversion works for construction of a dam, etc. The contractor may be required to quote a price for following a design shown in the contract drawings, but be permitted to offer an alternative design of his own. The option has to be made fully clear in the contract documents, and full details of the tenderer's alternative design have to be provided.

Checking tenders

Detailed consideration of tenders will usually start with an arithmetic check of the lowest three or four offers which are free of unacceptable qualifications. Any arithmetic errors found should be dealt with in a manner which is set out in the Instructions to Tenderers. Usually the unit rates quoted for items are taken as correct, and all consequent multiplications and additions are arithmetically corrected. Where this results in an alteration to the total tendered sum, either this altered sum is adopted, or the altered sum is brought back to the quoted tendered sum by inserting an 'Adjusting item' – the Instructions must state which. A reason for leaving the quoted tender sum unaltered is that the contractor's estimating staff work out the unit rates, whereas the contractor's directors will not check a tender arithmetically but will look at the total sum tendered to decide finally whether or not it is sufficient to cover the whole job.

The lowest three or four tenders are then checked for compliance with contract and other requirements, under the following headings:

- *Compliance*: conformity with instructions; completeness of entries; compliance with bond and insurance; absence of unacceptable qualifications, etc.;
- *Technical*: conformity with specification; proposals for materials; use of sub-contractors; temporary works proposed and methods of construction; intended programme, etc.;
- *Organizational*: staff proposed, experience, responsibilities held, etc.;
- *Financial*: make-up of total price; amounts for items in Preliminaries; unit rates; exceptional prices, errors and omissions, etc.

In addition, if open tendering has been adopted, details of tenderers' resources, past experience, financial and other data will need to be examined. This work will be conducted in parallel with the checking of prices described below.

6.7 Checking prices and comparing tenders

The engineer or consultant advising the employer must bear in mind that his report on tenders should provide the factual results of his analysis of tenders. He may need to indicate what any particular finding implies; but he does not recommend which tender should be accepted unless the employer requests this. Even so the choice of contractor must be the employer's, and not his advisor's. Sometimes it is necessary for the engineer or consultant to present an interim report on tenders. This can occur if there are many tenders, or complex issues need to be resolved concerning qualifications attached to tenders, or relating to the standing of tenderers if open tendering has been adopted.

While the lowest total tendered sum may be a major factor influencing choice, individual rates and prices must be examined to see whether relatively high or low rates entered could alter the ranking of tenders should certain quantities under re-measurement for payment come different from those in the bill of quantities. A contractor is entitled to set highly profitable rates for some items and non-profitable or loss rates for others. This can lead to problems if quantities are not as billed, or work has to be varied. The implication of such differences needs to be considered. Nevertheless prices for the same type of work can vary widely from one contractor to another; in this connection sums entered by the contractor in the Preliminaries Bill must be taken into account (see Section 15.10). One tenderer may put large sums there for access, insurance, setting up, etc.; another may spread the cost of such items over all his unit rates entering only a few, relatively small sums in the Preliminaries Bill. Differences in rates can also arise from different materials or methods used, different appreciation of risk, and sometimes from simple error.

If the lowest tender appears impracticably low, the employer may agree that the engineer should interview the tenderer in the hope of elucidating whether this results from the tenderer's inexperience, over-optimism, or misunderstanding of the contract requirements. However, such a meeting can prove uninformative leaving the problem still open as to whether such a tender should be accepted. Acceptance of a tender which would put the contractor to a certain loss can lead to skimped work or the contractor failing to complete the works. To allow the tenderer to adjust his faulty price would not be permissible for a public authority but he can be allowed to withdraw his offer. However, a private employer is not precluded from bargaining with a tenderer to settle an adjusted price, or to agree upon some other solution such as offering a bonus to make up the underpriced item if the contractor completes the works early.

The chances of receiving an unrealistically low tender can be minimized by avoiding open tendering and giving selected pre-qualified tenderers adequate time to prepare bids. Before tenders are received the engineer can estimate what a fair bid price should be. However, under fiercely competitive conditions lower bids may be received; or if there is much work available or the risks imposed on the contractor are high, bids can come much higher than expected. If a contractor expects he will meet administrative problems, difficulty in getting permits, payments, materials, consents, etc. and suffer from indecision or overcomplicated authorizing systems run by the employer, he will add a premium to his prices. It must be realized that contractors pay as much attention to the competence of employers, as employers pay to the competence of contractors.

A further matter to be examined is the effect of a tenderer's pricing on the rate of payments to him during construction, that is, on the cash flow. A contractor may set his rates for early work high, such as rates for excavation and foundation concrete. Thus these, containing a large element of profit to him, will provide him with a good inflow of surplus cash at an early stage in the project. Similarly he may enter high prices in the Preliminaries Bill for early temporary works, such as provisions of offices, etc. This pricing is of considerable financial benefit to a contractor, quickly reducing his start-up costs and borrowing needs; but it is also a dis-benefit to the employer who, often needing to borrow money to finance the capital expenditure on the project, has to pay interest thereon. Comparison of the rates of cash flow implied by different tenders may therefore need to be made to see their different financial effect on the employer. If the interest on a employer's borrowings is capitalized, that is, not paid when due but added to his borrowings, this can magnify the effect of early cash disbursement on the employer's costs, increasing the capital cost of the project to him. A further point is that a contractor who receives early money leaves the employer at extra risk, because if the contractor gets into financial difficulties, much of the early money may not have been spent on permanent works of value to the employer.

6.8 Choosing a tender

In order to resolve any mistakes or qualifications in tenders it is often necessary to hold a discussion with one or two of the tenderers. Such discussion must only take place with the knowledge of the employer, and in accordance with any restrictions set by him. At this stage there is a strong possibility that tenderers will know the prices quoted by at least some others, so the negotiator must be alert to any attempt by a tenderer to adjust his price, so as to become the lowest or, if he is already the lowest, to reduce the gap between himself and the next lowest. A contractor has admitted that 'It is always a great advantage to have the opportunity of taking a second look at one's tender when discussing it with a client.'

There has been a growing demand that tenders should be assessed on the basis of quality as well as price. Any quality criteria to be applied must be stated in the documents. Quality assessment is of major importance under design and construct contracts where the contractor has control of design, the materials supplied, and the workmanship. For such contracts the parties should endeavour to agree all significant design matters prior to award. For construction-only contracts the quality of the permanent works is already defined in the drawings and specification accompanying the tender documents, so it is the contractor's prospective performance that remains to be judged. This cannot be defined in measurable terms, hence the need for pre-qualification or similar assessment of tenderers.

The final assessment of tenders must, of course, take into account any rules set out in the tender documents. The EC Directives, for instance, require the methods of comparing tenders to be described in the documents, and they must be adhered to. Failure to do so could entitle a disappointed tenderer to challenge the award of contract to another and claim compensation for the lost opportunity.

Having made all necessary comparisons of tenders, a decision must be made as to which tender should be recommended for acceptance, if the employer requires this. Once tenders have been compared on a uniform basis and all non-conformities and queries have been resolved, the question must be asked: is there any cogent reason for not recommending the lowest ranked tenderer by price? Any reason put forward for this must be a real reason, such as clear evidence of a tenderer's financial problems or his lack of experience in some essential operation required under the contract. The report on tenders must be careful to present a balanced view, and not over or understate the case for any tenderer. It is important not to 'mix fact with opinion'. The facts or evidence should form one statement; any comment thereon should form a separate statement. Supporting information such as bankers' reports or references should be appended to the report. These problems of reportage are more likely to occur with open tendering. Under selective tendering, the competence of tenderers will already have been approved, so it is almost axiomatic that the lowest unqualified offer will be favoured.

6.9 Offer by a tenderer to complete early

A tenderer may state in his offer that his prices are dependent on being permitted to complete the works in a shorter time than the period for completion stated in the contract documents. This offer must be looked at with care because it implies that other separate contracts the employer may have let for supply of plant to be incorporated in the works must be speeded up also. Similarly any nominated sub-contractors must deliver earlier, and the engineer must be able to provide all outstanding design details according to the shorter programme. It is, of course, a benefit to an employer to have his works completed earlier: it can reduce his capital borrowing charges and enable him to gain an income from the works output earlier – though he must be able to accelerate his payments to the contractor. The question that arises, however, is whether the contractor's shorter time period should be substituted for the period for completion stated in the contract.

It is true that speedy construction can maximize a contractor's profit or permit him to offer a lower price, but this need not be his only motive. A contractor may say he can complete a project 3 months early if he suspects the job is so liable to delay by other contractors, nominated sub-contractors, extras, incompleteness of designs or unforeseen conditions, that he runs little risk of having to abide by his promise and indeed may be able to claim extra payment for any delay caused to him.

Therefore adoption of the contractor's time as the contract period for completion needs careful consideration and the position must be resolved clearly before award of the contract. The contractor might have second thoughts about his offer because he would become liable to liquidated damages if he did not complete in the time he offered.

6.10 Procedure for accepting a tender

After the closing date for tenders, and if tenders have not been publicly opened, contractors will be anxious to discover where they stand: either to prepare themselves for holding discussions over their tender, or to divert their energies elsewhere if they find themselves unlikely to be offered the contract. If prices have not been arithmetically checked, it is inadvisable to give any information lest it turn out misleading. However, when the ranking of tenders has been checked, it should be possible to inform contractors enquiring if they are unlikely to succeed. Once a decision has been made by the employer, all tenderers should be informed by a standard letter, stating the prices received but not identifying the tenderers who submitted them.

A **valid contract** must incorporate three basic elements:

- an offer (e.g. the tender) and its acceptance;
- consideration (i.e. the contractor undertakes to construct the works and the employer undertakes to pay him for them);

- an intent that the contract be legally binding (as evidenced in the contract documents).

During any negotiations the original tender may have been amended by interchange of letters. These letters must make clear what is the final amended tender offered and accepted. If the correspondence is not complete and some condition or qualification remains unsettled, then a contract should not be formed. So a check must be applied to ensure that everything has been settled. Once this has been done and full agreement has been reached, then actual acceptance of a tender can take place.

In the case where an employer is a person or private company the employer can accept a tender by writing little more than 'I accept your offer'. However, some corporations, and most statutory or other authorities, may be required by their constitution or standing rules to enter contracts above a certain value only by a **deed** or formal agreement which has to be signed by an authorized person acting on behalf of the authority. Some authorities require the agreement to be **under seal,** that is, stamped with the corporate seal of the authority. Under any method, the acceptance must make clear what documents form the basis of contract.

Where a local authority can only enter a contract by means of a formal Agreement, a typical letter from the clerk of the authority to the contractor might be written as follows:

Dear Sirs,

Contract No. 64 for XYZ Scheme

I am pleased to inform you that my Council has resolved to accept your tender dated 4 January 1994 for the above contract. The contract will comprise the following documents:

1. Volumes 1 and 2 of the printed documents containing the Tender, Conditions of Contract, Specification, Schedules and Bills of Quantities all as completed by you.
2. Contract Drawings Nos. 1–45.
3. Tender Amendments Nos. 1–3 inclusive.
4. The following communications:
   (a) Your covering letter of 4 January 1994 with enclosures 1–5 inclusive;
   (b) Our letter to you of 10 January 1994;
   (c) Your letter of 12 January 1994;
   (d) This letter of acceptance.

The amended tender total is £123 456.00 as set out on the attached sheet.

A formal Agreement is being prepared and I shall be glad if you will advise me when you can call in to sign it.

Please now produce your Performance Bond and evidence of insurances as required under the contract.

Yours faithfully,

Clerk to the Authority

If an authority empowers its chief executive to accept a tender on its behalf, or a company allows its director to accept a tender, the letter can be a direct acceptance. However, if the authority or company is employing a consulting engineer to correspond with tenderers, then the consulting engineer usually has no authority to accept a tender, so he can only advise a tenderer that the authority or company have decided to accept his tender in terms similar to the above, or perhaps in the form 'On behalf of the ...' or 'I am instructed by the ... Council to inform you that your tender is accepted, etc.'

Where acceptance of a tender is not possible for some time, for example, because it requires agreement from government or from some international funding agency, etc., a 'Letter of Intent' can be issued by the employer. This states the employer's intention to sign a contract and may therefore request the contractor to start on some aspect of the work. The Letter of Intent must state what work can be started, and how and what payment will be made for such work should the contract not be signed. There will also be a clause which provides for the Letter of Intent to become void upon signing of the contract. The contractor has to respond accepting the terms of the Letter of Intent. Usually the matter is discussed prior, so that the terms of the Letter of Intent are agreed before it is written. However, a Letter of Intent can prove full of legal pitfalls should anything go wrong, so it is best avoided. It can be useful, however, for authorizing a plant supply contractor to start producing designs and drawings of equipment he is to supply, that is, work which saves time but involves no large financial commitment.

A tender needs to be accepted within a reasonable time of its submission, otherwise a contractor may have grounds for withdrawing it. Sometimes the employer stipulates for how long tenders are to remain open for acceptance, or a tenderer may state this in his offer. A contractor is put in a difficult position when there is an unexpected delay in accepting his offer because, although he does not wish to lose a job, the delay can cause his costs to rise if prices are inflating or work he hoped to undertake in two summers and a winter is delayed to take place during two winters and a summer.

Publications giving guidance on tendering

Tendering for civil engineering contracts. ICE, 2000.
Tendering procedure: procedure for obtaining and evaluating tenders for civil engineering contracts. FIDIC, 1982.
Standard pre-qualification form for contractors. FIDIC.

Appendix: UK Regulations

UK Statutory Instruments implementing EC Directives:

The Public Works Contracts Regulations – SI 1991/2680 (implementing EC Directive 93/37/EEC) as amended by SI 2000/2009*

The Public Services Contracts Regulations – SI 1993/3228 (implementing EC Directive 92/50/EEC) as amended by SI 2000/2009*

The Public Supply Contracts Regulations – SI 1995/201 (implementing EC Directive 93/36/EEC) as amended by SI 2000/2009*

The Utilities Contracts Regulations – SI 1996/2991 (implementing EC Directive 93/38/EEC)

Note*: **SI 2000/2009 is **The Public Contracts (Works, Services and Supply) (Amendment) Regulations** which implements EC Directive 97/52/EC, amending inter alia – lists of contracting authorities, financial thresholds, time limits for receipt of tenders, and forms of model notices. It permits submission of tenders by electronic means.

7

The contractor's site organization

7.1 Contractor's site personnel

The key personnel employed by a civil engineering contractor on a construction site are usually:

- the agent, who is in charge;
- sub-agents and/or section engineers;
- the plant manager;
- the general foreman;
- a quantity surveyor or measurement engineer;
- the office manager.

On large complex jobs there may be several sub-agents or section engineers, each responsible to the agent for some part of the construction. The plant manager or 'site co-ordinator' organizes all plant required on the job, including its maintenance and any repair that can be done on the job. The general foreman is usually a widely experienced 'outside man' whose main job is to organize and direct the work of the tradesmen and the skilled workers on site. He will work closely with the sub-agents and usually have section foremen working under him.

The quantity surveyor will prepare the contractor's accounts, using the sub-agents or section engineers to supply him with the measurements of work done. The office manager will have an ordering clerk who issues orders for materials and gets invoices checked; and a pay clerk who checks the time sheets, makes up the pay sheet, and pays the men. On the smaller civil engineering jobs there will often be only an agent, a site engineer, a general foreman and an office manager.

In the head office of the contractor will be a contracts manager (sometimes a director of the firm) who is responsible for head office services to the job and who decides overall policy for it. He may advise on technical problems in the execution of the contract, but he does not direct its day-to-day execution. He may frequently visit the site but he is not full time on site. In the larger contracting organizations which have many projects in hand, there may be – commercial managers, project or contract managers, a chief engineer with engineering staff, quantity surveyors and estimators, and a safety manager.

Men and women are employed in any of the foregoing positions in the UK and other countries.

7.2 The agent

The agent is responsible for directing the construction work on site. He (or she) will have wide powers to employ men, hire machinery and equipment, purchase materials, and employ sub-contractors. His powers to do this without reference to his head office, will depend on the size of the job, its nature and distance from head office (particularly for overseas work), and his standing within his firm. The agent must be knowledgeable in the arts of construction, able to command men and be a good organizer. He needs a sound business sense, because his job is not only to get the works built properly in accordance with the contract but also to make a profit for the contractor. Some agents have risen to their position mainly by experience gained through many years on construction, others are professionally qualified engineers. A good agent is probably the most secure guarantee an employer can have that his works will be built well.

Control of the work is exercised through 'down the line management' which operates through a hierarchy of responsibility. Directions proceed from the agent, through his sub-agents, to the foremen and then to the tradesmen and labourers. This is necessary so that each person is clear as to what his responsibilities are and what he is supposed to do. Thus, if the agent sees some work being done which does not meet his approval, he will issue his instructions via the sub-agent in charge of that work. Day-to-day instructions are usually given verbally; they need to be clear, as simple as possible, and not capable of misunderstanding. This is not always easy to achieve when complex situations arise. Unnecessary explanations accompanying an instruction are best avoided because this can sometimes lead to misunderstandings due to pressure of work.

The agent's chief problem is to keep the work progressing as efficiently as possible. His main troubles occur when an unexpected difficulty is encountered, or there are problems with labour, plant, or materials. When any of such difficulties occur, the agent may have to change the day's plan of work and issue new instructions. He has to choose between the options open to him, bearing in mind both his short-term strategy for the next few days, and also his medium-term strategy of what operations must be completed within the next 2 or 3 weeks. As in a game of chess, present moves have to be decided in terms of some overall strategy, the moves having to be re-thought whenever circumstances change – especially the weather.

7.3 Site field personnel

Section engineers carry out or organize the surveying and setting out work, and conduct any necessary technical tests. Initially there will be considerable work to do in site levelling, and setting out the main grid lines for the project. There will then be much detailed setting-out work to do, as required by the foremen on the works. Temporary works may have to be designed and set out, such as access roads, power lines, water supply lines, drainage, concrete foundations for the batching plant and cranes, and so on. In addition it is normally the job of the section engineer to record progress and keep progress charts up-to-date. On small sites, the job of sub-agent and section engineer may be combined.

The plant manager holds a key position on site. His job can be onerous since construction work is held up if plant is not available due to breakdowns or failure to order in time. For sites in the UK and other developed countries much of the plant used on site is hired and kept in maintenance by the hirer. This requires constant liaison between the plant manager and the hire firms used. Where the contractor's own plant is used, maintenance and repair of this will be needed. Assisting the plant manager will be fitters and welders and he will often have to get repairs done at times outside working hours when construction is not proceeding. He will also have to maintain power supplies to the site and its offices.

A *general foreman* is widely employed on the many construction projects which are not too large for one person to control. He then acts as the agent's right hand man for the execution of the work in the field, his duty being to keep the work moving ahead daily as the agent has planned it. He often has much authority on site, and any junior engineer who gets at cross purposes with him may find his days numbered. Such men are often astonishingly capable from their long experience of construction. For instance, their familiarity with soil characteristics may often enable them to judge by eye that some foundation or fill material is 'no good', long before a site engineer's tests prove it so. He will have a knowledge of what machines can do, and the basic principles of surveying and levelling. At his best he is an all-round craftsman in the art of civil engineering construction, and many of the great constructions of the past owe their quality to the general foreman who took charge of their construction. The professional engineer can often learn much from him. On many civil engineering jobs the general foreman is the key outside person in charge of construction.

The *skilled men* include reinforcement fixers, steel erectors, concreters, formwork carpenters, bricklayers, pipe jointers, crane and machine operators, miners and other trade specialists. The contractor will often have a small nucleus of experienced tradesmen in his permanent employment, getting additional tradesmen through the local employment office, or advertising for them. *Specialist sub-contractors* or *labour-only gangs* are now widely used to carry out specific trade work. Labour-only gangs are self-organizing groups of workers under their own foreman or gang leader. Quite often travelling gangs of formwork carpenters, steel erectors or reinforcement fixers are taken on. Hand excavation of tunnels was almost always undertaken by an experienced gang under a leader, because the work demands close teamwork. Once a gang proves its worth, an agent will endeavour to use the same gang on his next job if he has similar work to do. Such gangs of tunnellers, formwork carpenters, or steel fixers are employed as a whole, so any unsettled dispute arising between the gang leader and the agent – usually about pay or conditions – may lead to the gang leaving en bloc bringing the job to a standstill.

On overseas jobs in the less developed countries much manual labour is still used, not only because of low rates of pay and the cost or difficulty of getting machinery, but because it is the traditional way of undertaking construction which suits the local economy and workpeople. In some countries women are widely used to undertake manual labour. If machines are brought in to do most of the work, this can deprive the local economy of a benefit. For projects in underdeveloped countries, an international funding agency will often require that as much use as possible is made of local labour to reduce offshore costs. It is important to recognize that this inexperienced labour may require tuition before they can be expected to reach an acceptable level of output. Also provision of adequate living conditions and canteen services, plus training in safety, may be essential to improve the well being and output of such employees.

7.4 Site office personnel

An *office manager* is needed on all but the smallest sites. He deals with getting all the miscellaneous requirements for the job, that is, the 'consumables' such as picks and shovels, protective clothing, small tools, minor repairs, fuel deliveries, electricity supplies and telephone, etc. He will be in control of storekeepers, messengers, teaboys, staff car drivers and night watchmen. On small projects he may order materials for the construction, as instructed by the agent, so will have to deal with the invoices for such materials, checking invoices against materials delivered, signing and sending them to head office for payment. On larger sites he will have an *ordering clerk* to do this for him.

A *site accountant*, often assisted by a *pay clerk*, handles all cash transactions on site and the local bank account. It is essential to employ experienced persons on this type of work. Even taking the sealed pay packets around to the workers is best done by an experienced pay clerk who knows what care is needed to avoid the upset which occurs if a pay packet 'goes missing'.

For the supply of materials in regular use, such as concrete aggregates, ready-mix concrete, cement, bricks, timber, etc. the agent will seek out local suppliers, get quotations from them and pass them to the head office buyer with recommendations. The head office buyer may then set up standard agreements with the local suppliers recommended by the agent, or he may discuss with the agent, use of some alternative supplier. Actual requisitions for delivery can then be placed by the agent direct with the supplier, with copies sent to head office. A *materials clerk* on site then checks the deliveries against the supplier's invoices and against the original order, certifies the invoice and sends it to head office for payment. This system can only work, of course, if both head office and the site are in the same country. Overseas, the agent has to carry out all the work involved, using sub-agents and accountants or other supporting staff to carry out the work for him.

A *quantity surveyor (QS) (or surveyors)* may be employed on site to draw up monthly applications for payments due to the contractor, according to the measurement of work done as required under an Institution of Civil Engineers (ICE) bill-of-quantities contract. Alternatively these QSs – as they are widely called – may be based in head office visiting site monthly. To them the sub-agents or section engineers submit their monthly measurements and the QSs then make up the monthly statement, including any claims the contractor makes for additional payment for extra work done, delays or difficulties encountered which the contractor considers should be met. However, on small jobs quantity surveyors are not always employed because civil engineering quantities differ from building quantities (see Sections 15.4–15.7), so the small contractor may use his engineer for this task or do it himself.

7.5 Accounting methods

While the agent may have wide authority on site, the large sums of money he commits his firm to spending must come under the control of the contractor's head office. For supply of materials in regular use the system described in the previous section is used, that is, head office places the contracts for supply as agreed with the agent and pays the suppliers' invoices for deliveries as checked by the agent. In this manner the agent does not have to handle large payments himself.

But the agent will also need to open a local bank account, into which head office transfer funds for a variety of cash payments – for fuel for vehicles, a variety of consumables and for the wages of labour taken on by the agent. These payments have to be handled by the site accountant and his pay clerk on site. The paysheets are made up by the pay clerk on the basis of time sheets sent in to him by the men on site, the section foremen or gangers certifying the timesheets of men working under them. From time to time an accountant from head office may visit the site to audit the local paysheets and cash disbursements by the agent.

On overseas projects the large sums of money which have to be expended locally mean that a fully staffed accountant's office, under the charge of an experienced site accountant, will need to be set up.

The monthly returns of local expenditure sent by the agent to head office are then added to all the payments and charges met by head office and debited to the job. These will include not only invoices paid for materials delivered to site, but also all other relevant expenditure, such as salaries of those working on the project, cost of any equipment purchased for it, hire charges for plant, insurances, etc. and such head office oncosts that the contractor currently applies to projects in hand. Head office should keep the agent informed of the figure of total expenditure to date; but this figure inevitably lags behind the actual expenditure *commitment* to date because of the time delay between ordering materials and entering payment for the same in the books. Hence a prudent agent may keep such an assessment going himself because of the importance of controlling the overall expenditure on the job. His system may not be exact, but his better knowledge of what expenditure is currently committed may give him a useful guide as to how the job is progressing financially.

Most accounting in a contractor's head office is now done by computer using codes for different sites and classes of expenditure. Such a system can be advantageous if it shows costs for different elements of a job or types of work, which can help in building up a record of unit costs which can act as a guide for future bids, or may be useful in formulating any claims. In practice, however, such systems seldom have sufficient definition for this purpose, but are predominantly used to show the current profit or loss on a job.

7.6 Providing constructional plant and equipment

A contractor will own a stock of plant and equipment which is available for loan to construction jobs the contractor has in hand. When items are loaned to site, the job account held in head office will be debited with the cost of plant delivery, plus rates per day (or per hour) according to whether the plant is working or standing idle on site. These rates are termed 'internal hire rates'. Plant not available from stock will need to be obtained by the agent from some outside plant hirer, who will charge 'outside hire rates' which are usually higher than internal hire rates. An agent may also choose to use plant from a local plant hirer because the cost of delivery may be less than that from the contractor's plant depot if the latter is remote from the site.

Internal hire rates for plant will need to cover the cost of plant depreciation, running maintenance, major overhauls and renewals, plant depot and administration costs and some adequate return on the capital investment involved. The cost of working repairs to plant is high, representing some 25 per cent or more of the normal commercial outside hire rate. The frequency of repairs is particularly high for mobile plant, where tracks may need frequent attention, and tyres may need renewal at high cost every few months. Wire ropes for cranes need constant renewal and a stock of same has to be kept on site.

Decision as to what plant and equipment should be owned by the contractor is a complex matter. Easily transportable equipment which can be used several times, such as temporary site offices, is commonly held in stock by a contractor. Plant with a long life and little maintenance, usable on many jobs – such as flat wheel diesel rollers – might also be held. But deciding what other major plant should be held for hiring out to sites involves many considerations such as how often will it be used; its cost and expected working life; transportability; insurance, running and maintenance costs, and whether the resulting estimated internal hire rate gives an adequate return on capital invested, and shows a worthwhile saving over outside hire rates.

The risks involved in using owned plant have to be taken into account also. Breakdown repairs of some plant can cause several days' delay to the whole job which may be very costly; whereas if hired plant breaks down it may take only a day or two to get a replacement from the hirer or from some other firm. Also hire firms can often supply an experienced driver with plant, and their hire charge will cover all maintenance, repair, breakdown and renewal costs, which reduces the contractor's on-site commitments and risks. Additionally some operations, such as bulk excavation, may be let to a sub-contractor who provides all plant and drivers required for payment of either a fixed lump sum or more often unit rates for the measure of work done.

7.7 The contractor's use of sub-contractors

Many civil engineering contractors now use sub-contractors to do much of their work. Most conditions of contract permit a contractor to sub-let work of a specialist nature; but the ICE conditions of contract have gone further and permit the contractor to sub-contract any part of the work (but not the whole of the work), subject only to notifying the engineer of the work sub-contracted and the name of the sub-contractor appointed to undertake it.

The contractor does not have to notify any labour-only sub-contracts he uses.

The engineer can object, with reasons, to the appointment of a sub-contractor, but otherwise has no rights in connection with such sub-contracts, except that he can require removal of a sub-contractor who proves incompetent or negligent, or does not conform to safety requirements. Under FIDIC conditions for overseas work, sub-contracting requires the engineer's prior sanction.

In building work there has long been a trend to pass the majority of work to sub-contractors who specialize in various trades, and the same has now occurred in civil engineering where many operations are 'packaged up' and sub-let. Thus sub-contracts may be let for excavation, formwork, reinforcement supplied and erected, and concreting. The advantage to the contractor is that this reduces the staff he needs on site and his capital outlay on plant and equipment. He can use sub-contractors with proven experience and does not have to take on a range of temporary labour whose quality may be variable. The contractor retains responsibility for the quality and correctness of work and, of course, has to plan and co-ordinate the sub-contract inputs, and often supply any necessary materials.

But if much of the work is sub-contracted, the contractor's or agent's main input to a project may be that of dealing with the sub-contracts and controlling their financial outcome, so these matters may take priority over dealing with any engineering problems which arise. The contractor may therefore tend to leave a sub-contractor to solve any problems he encounters, on the basis that these are his risks under his sub-contract and it is up to him to deal with them. But the sub-contractor may think otherwise, so a dispute arises as each considers the other responsible for any extra cost or delays caused.

Frequent disputes have also arisen in recent years when any default or presumed default by a sub-contractor has resulted in the contractor withholding payment to him. Late payment by contractors to sub-contractors is another widespread source of complaint by sub-contractors, but remedies are difficult to devise. The sub-contracts are private contracts whose terms are unknown to the engineer and the employer, so they cannot interfere in any such dispute. The engineer has only power to protect nominated sub-contractors, i.e. sub-contractors he directs the contractor to use (see Section 15.8).

7.8 Recent measures to alleviate sub-contract disputes

The problems between main and sub-contractors were one of the areas to benefit most from Part II of the UK Government's *Housing Grants, Construction and Regeneration Act 1996* (see Section 1.6). The introduction of adjudication under that act to deal with disputes has at least allowed sub-contractors to press their claims to an earlier conclusion, and to challenge any withholding of payment by the contractor.

The Act requires payment terms to be stated and regular payments made. It prohibits 'pay when paid' clauses, and requires the contractor to issue a detailed 'withholding notice' if he seeks to hold back payment. These measures have eased the cash flow problems of sub-contractors. Also most standard forms of sub-contract now contain provision for payment of interest on delayed payments, but this may not be very effective because a sub-contractor may not claim interest for fear the contractor might not as a consequence give him any further work.

The Civil Engineering Contractors Association (CECA) has issued a Form of Sub-contract 'for use in conjunction with the ICE conditions of contract.' Contractors are, of course, not obliged to use this form and many use one of their own devising or modify the standard form. The provisions of the CECA sub-contract illustrate the many matters which such a sub-contract has to cover and the difficulty of trying to provide rights to the sub-contractor without putting the main contractor at risk under his contract.

Provisions of the CECA sub-contract, apart from defining the work, timing and duration of the sub-contractor's input, require the sub-contract to set out the division of risks as between contractor and sub-contractor. It defines procedures and methods of valuing variations made by the engineer and confirmed by the contractor, or made by the contractor; and sets out procedures for notification and payment for 'unforeseen conditions' or other claim matters. It also stipulates requirements for insurances and so on. Many of the provisions are similar in terms to the ICE conditions applying to the contractor, and are thus passed on to the sub-contractor in respect of his work. The sub-contractor is 'deemed to have full knowledge of the provisions of the main contract' and the contractor must give him a copy of it (without the prices) if the sub-contractor requests it.

Of particular importance is Clause 3 of the CECA sub-contract which requires the sub-contractor to carry out his work so as to avoid causing a breach of the main contract by the contractor. He has to indemnify the contractor 'against all claims, demands, proceedings, damages, costs and expenses made against or incurred by the contractor by reason of any breach by the sub-contractor of the sub-contract.' But a sub-contractor undertaking a small value contract may find it impossible to accept this clause. If he fails to complete his work on time and this could possibly cause a delay to the whole project, he might be liable to pay many thousands of pounds to the contractor – far in excess of the value of his sub-contract.

A further problem for the engineer is that, if a dispute arises between the contractor and his sub-contractor as to who is responsible for some defective work, the defect can remain uncorrected until the dispute is resolved. If a defect is found after the sub-contractor has left site and he is believed or known to be responsible for it, the contractor may not be able to get the sub-contractor back to site to remedy the defect, or to pay for its repair. To guard against this, the contractor may therefore hold back full payment to the sub-contractor for many months until a certificate of completion for the whole works is issued. This will cause another dispute between contractor and sub-contractor.

The development of sub-contracting in civil engineering has therefore brought both advantages and disadvantages. However, problems rarely arise if the contractor can use sub-contractors he has worked with before whose work has proved satisfactory and he treats them fairly.

8

The employer and his engineer

8.1 Introduction

When the employer has drawings and specifications prepared there are two main types of construction contract he can use in the UK to get the works built – the ICE Conditions of Contract (the 'ICE conditions') or the ICE Engineering and Construction Contract (the 'ECC conditions'). These have been described in Sections 4.2(a) and 4.2(f).

The ICE conditions have been used for construction of works for many years, are comprehensive in their provisions, and are still the most widely used conditions. The ECC conditions for the construction of works are not so extensive and detailed as the ICE conditions, and the FIDIC conditions for construction of works overseas are very similar in terms to the ICE conditions.

Hence the provisions of the ICE conditions are fully described below, and any different provisions of the ECC or FIDIC conditions are noted in this chapter or later.

8.2 The role of the employer's engineer under ICE conditions

Under the ICE conditions the employer appoints an independent engineer to administer the contract for construction termed 'the Engineer' under the contract. This engineer is required under the ICE conditions to 'act impartially within the terms of the contract having regard to all the circumstances' (Clause 2(8)). He (or she) may often be a consulting engineer engaged by the employer, or can be a member of the employer's staff, but this does not affect the duty to act impartially.

The advantage of employing an engineer who has to administer the contract impartially is that both the employer and the contractor can expect their interests to be dealt with fairly. Also when the contractor can expect fair payment for extra work ordered or arising from some unforeseen trouble, his risks are reduced, thus enabling him to submit his keenest prices. Both the employer and the contractor can, however, challenge any decision of the engineer by taking the matter in dispute to a conciliation procedure, adjudication, or to arbitration for settlement.

Since the employer does not administer the contract he cannot issue an instruction direct to the contractor, he can only request the engineer to do so. But the engineer is bound by the terms of the contract, so if he finds he has no power to implement the employer's request, or thinks to do so would amount to an unfair administration of the contract, then the employer has to put his request direct to the contractor for settlement outside the terms of the contract. This rarely happens, but as an example, if the employer wants the contractor to stop working for a day so that he can bring a party of visitors on site to view the construction, he has to seek the contractor's agreement to this because the engineer usually has no power to order this.

The engineer's duties set out under the contract are extensive. Under the 7th (measurement version) of the ICE conditions these duties include the following:

- Clause 5: explaining any ambiguity in the contract documents.
- Clause 7: issuing any further drawings or details needed for construction.
- Clause 12: confirming or deciding on any actions to overcome unforeseen ground conditions should these be encountered.
- Clause 13: ensuring that the works are constructed in accordance with the contract.
- Clause 14: checking that the contractor's programme and his methods of constructing the works comply with any specified needs and permit the work to be finished without harm to the permanent structures.
- Clause 36: testing or witnessing tests on materials either during manufacture or on the site, and (Clause 38) examining any work such as foundations which will be covered as construction proceeds.
- Clause 41: fixing the date for commencement of the work and, (Clause 40) ordering suspension of the work or part of it if this proves necessary.
- Clause 44: determining any extensions to the time allowed for completion of the works and (Clause 48) certifying when completion has been achieved.
- Clause 51: ordering and (Clause 52) valuing variations to the works.
- Clause 52(4): keeping records of facts relating to any claims made by the contractor and deciding the amount, if any, of extra payments due as a result.
- Clauses 55–57: measuring and valuing the works constructed.
- Clause 60: considering the amounts of interim and final payments to the contractor and certifying those amounts as are in his opinion due.
- Clause 66: giving his decision on any disputes specifically referred to him; such decisions being subject to adjudication or arbitration if not accepted by the employer or the contractor.

8.3 A note on alternative provisions of the ECC conditions

Under the ECC conditions a project manager is appointed to administer the contract and he has no duty to act independently or impartially.[1] He represents the employer, so acts for the employer who is committed by his manager's decisions. Consequently the employer has no right under the contract to take a dispute with his manager to adjudication or arbitration. But if the contractor disputes any action of the project manager, this comprises a dispute between the contractor and the employer which can be taken to adjudication or arbitration.

A supervisor on site (with assistants if need be) is also appointed to carry out certain specified duties relating only to the quality of construction. He inspects and tests the work (Clauses 40 and 41) and instructs the contractor to search for and remedy defects (Clauses 42 and 43). He submits reports to the project manager and the contractor. Where his appointment is separate from that of the project manager, their respective responsibilities need to be carefully defined and co-ordinated.

The project manager's duties include many similar to those listed above for the engineer under the ICE conditions, in particular under the ECC these include:

- giving early warning of changes (Clause 16);
- resolving ambiguities in the documents (Clause 17);
- deciding and certifying completion (Clause 30);
- accepting or not accepting the contractor's programme (Clause 31);
- instructing a suspension of work (Clause 34);
- certifying take over of the works (Clause 35);
- assessing and certifying payments due (Clauses 50 and 51);
- deciding on compensation events, asking for quotations from the contractor for these and assessing any payment or time extension due (Clauses 60–65).

Further differences between the ECC conditions and ICE conditions are dealt with in Sections 17.3, 17.8, 17.11 and 17.12.

8.4 Limitations to the engineer's powers under ICE conditions

Under the ICE conditions the engineer can only instruct a variation of the works which is 'in his opinion necessary for the completion of the works', or 'desirable for the completion and/or improved functioning of the works'. Thus the engineer cannot order matters which are, for instance, extraneous to

[1]Although he has no duty to act impartially, he will in practice do so, to avoid a dispute arising which the contractor takes to adjudication or arbitration.

the works or which add entirely new items; these are matters the engineer must refer to the employer who will need to negotiate with the contractor his agreement to undertake the addition (see Section 17.3).

Although the engineer is given a wide range of powers, he should not use them without reference to the parties to the contract, either of whom may wish to state his view on matters the engineer has to decide. The FIDIC conditions, for instance (see Section 4.3), specifically call for such consultation by the engineer as part of the procedure he must adopt before arriving at his decision.

However, if the employer wishes to restrict the engineer's powers which would otherwise be exercisable under the contract, the employer must state in the tender documents the specific powers which the employer reserves for himself. Both the ICE conditions Clause 2(1)(b), and the FIDIC 4th edition conditions require this. But it is unwise for the employer to reserve too many powers for himself, because this could affect the basis of contract and reduce the benefit of having an independent engineer. Tenderers might then take a different attitude towards the contract, since a tenderer may only offer his lowest price if he is confident that an independent engineer will administer the contract. Employers should also be aware that prior approving of matters such as extension of time or claims may restrict their ability to dispute them later.

However an employer may sometimes wish to ensure that he is involved in decisions likely to cause additional expenditure above some given limit, or which alter significantly some aspect of the works. In practice, such restrictions are unlikely to detract from the engineer's independent position because the engineer should keep the employer advised of such matters and endeavour to agree with him what should be done. Most extra costs arise from having to deal with unforeseen conditions which must necessarily be dealt with, or from alterations required by the employer himself.

A different situation can arise if it becomes evident that the estimated final cost of the contract is approaching or likely to exceed the contract sum. In that case the engineer must forewarn the employer in good time, because an employer such as a government or local government authority, may have no authority to spend more than the contract sum, or may need to go through a lengthy procedure to obtain sanction for any excess expenditure. In these circumstances the employer may need to step in and negotiate with the contractor a change to the works required, or perhaps deferment of construction of part of the works to some later date.

8.5 The engineer's duty to provide all necessary drawings to the contractor

Under the ICE conditions the engineer has a duty to provide the contractor with the drawings and further instructions needed to carry out the works. This is additional to the tender drawings issued which do not need to show every detail. The engineer must therefore watch construction progress to ensure any further drawings the contractor needs are supplied to him in good time. These may include drawings from plant suppliers of the foundations required for their plant and so on. If the engineer does not supply such drawings in time, the construction could be delayed, causing the contractor to claim for delay to part or whole of the job and any extra cost arising, which will have to be met.

If the design of the works (or part of them) has not been undertaken by the engineer for the construction but by some other firm, the engineer will have to ensure they produce any further drawings and information required in good time. The engineer then has less control over the situation, with a greater chance of delays and errors arising. Time must be allowed for the engineer's checking and possible amendment of designs submitted by others. A prudent engineer will ensure that all such information is in his hands as soon as the construction contract has been let.

The engineer may require the contractor to supply drawings and details for his temporary works, such as formwork, including design calculations for the same. These must be checked and consented to by the engineer to ensure they are suitable and not detrimental to the permanent works. Time must be allowed for this process including time for any possible amendments.

Designs will also have to be checked against safety requirements of the Construction, Design and Management (CDM) Regulations (see Sections 10.2 and 10.3).

On some large jobs or those overseas, there has been a practice to divorce construction from design. The employer uses one firm to produce design drawings and specifications, on receipt of which the employer pays the designer off. The employer then uses the drawings and specifications to get tenders for construction, and engages another firm to supervise the work of the construction contractor. This approach can be very unsatisfactory because, if constructional difficulties are encountered or variations prove necessary, the measures taken may not be in line with design assumptions made by the designer. The firm supervising construction will have no rights to contact the designer, and the designer has no obligation to provide any further information.

For some types of structures, such as dams or earthworks, where the safety and durability of the structure is highly dependent upon the nature of the foundations and materials used in the construction, a responsible engineer or firm of consultants would not be prepared to undertake the design without also having rights to supervise construction.

8.6 Quality assurance considerations

A contractor may run a quality assurance (QA) system and some employers take this into consideration when making a list of selected contractors for tendering. QA is an administrative system for checking that the quality of a firm's output complies with some set standards (see Reference 1). But this does not include a definition of the standards. For example a contractor may issue a design manual for formwork; this is his 'quality standard'. His QA system then only stipulates the actions required to ensure conformity to such standards. Such actions may include:

(a) designers must use the design manual;
(b) must have their designs checked by the firm's formwork specialist;
(c) the specialist must check and sign the design as approved;
(d) the signed design sheets must be filed, indexed and kept;
(e) the agent or his site engineer must check and sign that the formwork is erected as designed;
(f) the contractor's safety supervisor is to inspect and sign that the formwork erected is safe for use.

A QA system can cover a few or a whole range of a firm's operations, but to ensure that it meets the intended objectives (which have to be defined) it has to be **audited**. Audits can be carried out internally by a member of the firm, or by a client proposing to employ the firm, or by an independent authorized certifying body who can issue a certificate of approval (see Reference 2). In the last case the QA system is said to be **certified**. Repeat auditing is required from time to time.

A supplier may say he runs a QA scheme to the current standard of ISO 9000, but this has nothing to do with the standards he adopts for his products which need not conform to any quality standard. Also a contractor can have a QA system but people may fail to follow it. A QA manager can be appointed to see the system is operated; but he will not know when checkers have signed without actually checking, nor may he know when checks have been missed.

On site a QA system can be difficult to run because most instructions will be given verbally, checks are visual, and much work is sub-contracted or done by temporary labour. Thus a QA system can exist, but it may not be effective. A 1994 report on seven major projects for the UK Concrete Society gave many instances of defects observed in concrete design and construction despite QA systems being run (see Reference 3).

A further difficulty is that the engineer must have the contractor's QA system checked. This involves ensuring that

(i) check procedures cover all necessary elements of work;
(ii) the procedures have a reasonable chance of providing the standards required;
(iii) auditing such procedures on an irregular and selective basis gives assurance they work effectively.

Setting up QA procedures and auditing are specialized activities and many companies have staff trained to carry out these tasks. A useful publication is the *ISO 9000 pocket guide* (see Reference 4).

There has been extensive debate as to whether a contractor's QA scheme could permit reduction of the engineer's role in supervising the contract for construction; the idea being that the resident engineer would then only need to check that the contractor's QA checking system was being properly applied (see References 5 and 6). But under ICE conditions the engineer has a responsibility for ensuring the quality of the work is as specified and he cannot pass this duty to others. Even under ICE design and construct conditions and the ECC conditions the employer's manager has powers, and therefore implied duties, to ensure work is satisfactory or defect-free. Such contracts would have to be radically re-worded if sole reliance were to be placed on a contractor's QA system for quality of work done.

Only under some kind of turnkey or simple purchase contract might reliance be placed on a contractor's QA system though, even with that type of contract, an employer may often appoint an inspector to watch over the contractor's work on his behalf. The presence of a good inspector gives the employer, his engineer, and the contractor assurance that the work is inspected and is satisfactory. His cost to the job may be no more than the increased price a contractor might charge for running and auditing a QA system and the cost to the employer of having to check the contractor's QA system, and may give a better guarantee of satisfactory workmanship.

References

1. BS 5750 Part 1:1987 *Quality systems: specification for design/development, production, installation and servicing* (equivalent standards are ISO 9001:1987 and EN 29001:1987).
2. The certifying bodies are monitored by the National Accreditation Council of Certifying Bodies (NACCB).
3. When quality takes a dive. *Construction News*, 26 May 1994.
4. David Hoyle, *ISO 9000 pocket guide*. Butterworth-Heinemann, 2000.
5. CIRIA calls for slashing of engineer's role to aid QA. *New Civil Engineer*, 11 June 1992.
6. Systems analysis. *Water and Environmental Management*, October 1993.

9

The resident engineer's duties

9.1 The engineer's representative on site – the resident engineer

The ICE conditions permit the engineer to appoint an 'Engineer's Representative' on site, commonly termed the **resident engineer** to 'watch and supervise the construction and completion of the Works' (Clause 2(3)). The engineer can delegate to the resident engineer 'any of the duties and authorities vested in the Engineer' (Clause 2(4)) with certain exceptions, which are dealt with in Section 9.2.

The resident engineer therefore has to act at all times under the direction of the engineer, exercising only the powers delegated to him, impartially as the engineer is required to act. He must be aware that his actions commit the engineer, and therefore in all cases of doubt as to a proposed action, he should first report to the engineer. He may make suggestions to the engineer, point out difficulties and advise on their overcoming; and because he is full time on site he should be able to forewarn the engineer of problems lying ahead. In taking decisions he must be aware of his own technical limitations and always refer matters to the engineer which should be put in the hands of specialists or those more qualified to take a decision than himself.

The ICE conditions require the name of the person appointed as resident engineer (i.e. Engineer's Representative) to be notified to the contractor (Clause 2(3)).

The powers which the resident engineer can exercise on behalf of the engineer must be stated in writing to the resident engineer and copied to the contractor (Clause 2(4)).

9.2 Powers not delegated to the resident engineer

There are certain powers which the ICE conditions do *not* permit the engineer to delegate to his resident engineer. These are:

- payment or extension of time for adverse physical conditions or artificial obstructions (i.e. Clause 12 claims);

- extensions of time for completion;
- issue of substantial completion certificates, defects correction certificate and final certificate for payment;
- notice that the contractor has abandoned or appears unable to complete the contract;
- decisions on matters of dissatisfaction prior to adjudication or arbitration.

In addition it is often the case that the engineer does not delegate to his resident engineer in the UK power to:

- issue variation orders (VOs) or authorize payment to the contractor for delay;
- issue interim payment certificates;
- approve the contractor's programme for construction.

The purpose of the first of the last three exclusions is to permit the engineer (or staff acting on his behalf) to check both the justification and the amount payable under a proposed VO. However, if the site of construction is overseas the resident engineer may also be given powers to issue VOs and interim payment certificates. In this case the resident engineer would normally have appropriate staff on site, to check proposed VOs and interim payment certificates before issue.

It is to be noted that the FIDIC conditions for overseas construction do not restrict the powers the engineer can delegate to the resident engineer.

9.3 Usual powers delegated to the resident engineer

The usual powers and duties delegated to the resident engineer under the ICE conditions may contain most or all of the following:

- Agreeing details of methods of construction; checking that appropriate instructions are given and any information required by the contractor is supplied in good time.
- Ensuring that all materials and items to be supplied by the employer under other contracts which are to be incorporated in the works are ordered in good time.
- Checking that materials and workmanship are satisfactory and as specified; issuing instructions for remedying faults therein.
- Checking lines, levels, layout, etc. of the works to ensure conformity with the drawings.
- Issuing further instructions, drawings and clarifications of detail as are necessary to ensure satisfactory construction of the works.
- Measuring the amount of work done, checking the contractor's interim statements and preparing them for submission to the engineer.
- Undertaking all tests required and keeping records thereof.
- Recording progress in detail; keeping a check on the estimated final total cost of the project.

- Examining all claims from the contractor, preparing data relevant to such claims, sending to the contractor an initial response to every such claim.
- Reviewing dayworks sheets, increase of prices, and all other matters requiring accountancy checking.
- Checking the design of contractor's temporary works for compliance with safety regulations and satisfactory construction of permanent works.
- Acting as the engineer's Safety Supervisor on site (see Section 9.6).
- Reporting on all the foregoing to the engineer in the form he requires.

9.4 Some common problems

The following are typical of some common problems the resident engineer – frequently referred to as 'the RE' below – may have to deal with.

1. The contractor will not undertake some variation of the work the RE orders unless he receives a VO signed by the engineer in advance. This usually means the contractor wants to know in advance what he will be paid for the varied work. In most cases the RE should have enough experience to advise how the varied work will be paid for. The contractor is, however obliged to carry out such work as instructed and the engineer will have to issue a VO stating the pay rates to be applied as determined by the contract.
2. The contractor claims that the RE's clarification of the details of some work varies that work and entitles him to extra payment, but the RE decides no payment should be made. If the contractor continues in his claim, the RE should forward it to the engineer with details, forewarning the contractor that if the engineer agrees no payment is due, the contractor will have to decide if he will take the matter to adjudication.
3. The RE approves some material, workmanship or method of working of the contractor, but later finds the engineer thinks the RE's approval was wrong. If the error is one which the engineer feels he must rectify, he must do so by negotiation with the contractor, if necessary agreeing some extra payment to the contractor for abortive work. But, if the engineer suspects the contractor has taken advantage of the RE's failure to appreciate the consequences of the contractor's proposal, or has concealed such consequences from the RE, he may decide to countermand the resident RE's decision without agreeing any recompense to the contractor.
4. A problem frequently occurring is when the RE has to consider whether some excavation for a foundation has reached satisfactory foundation material. The problem can be compounded by the fact that to refer it to the engineer may cause a delay to the job, giving the contractor reason to make a delay claim. But, if the RE is in doubt as to whether the material is satisfactory he *must* refer the matter to the engineer (or his geotechnical adviser). However, the RE should have foreseen the problem and taken steps in good time to investigate what the ground conditions are likely to be, either by

undertaking hand augering of the foundation site or, if necessary, using the contingency money under the contract to instruct the contractor to excavate a trial hole paid under dayworks (see Sections 13.8 and 16.3).

9.5 Some important points the resident engineer should watch

Some important provisions of the ICE conditions of contract that need to be borne in mind by the RE are as follows:

1. All instructions to the contractor have to be given in writing or, if given orally, have to be confirmed in writing 'as soon as is possible under the circumstances' (Clause 2(6)(b)).
2. If the contractor receives an oral instruction and confirms it in writing, and the engineer does not contradict such confirmation 'forthwith', then the confirmation is 'deemed an instruction in writing by the engineer' (Clause 2(6)(b)). These confirmations of verbal instructions – or 'CVIs' as they are called – can raise special difficulties for the RE and the problems of handling them are dealt with in detail in Sections 13.3 and 17.6.
3. Although an RE may not have been delegated powers to decide how much should be paid (if anything) against a contractor's claim for extra payment, he has powers to write to the contractor stating his views on the claim. It is imperative he should do so, in each case, so that the facts as he sees them are on record.
4. There are numerous 'time clauses' in the conditions of contract, that is, clauses stipulating some time limit within which the engineer (and therefore probably the RE also) must take action. An important instance is the requirement that the engineer must comment on the contractor's proposed programme within 21 days of its receipt, otherwise the engineer is deemed to have accepted it (Clause 14(2)). The same, in effect, applies to any part-programme or revised programme the contractor supplies. Consequently if the engineer fails to comment within 21 days, the contractor's programme is deemed approved and anomalies may be introduced if the programme does not reflect the specified timing.
5. The RE has to ensure that the contractor receives all approvals, drawings, details and other information he needs to construct the works, in good time; otherwise the contractor may claim for delay (Clause 7(4)).
6. The RE should not accept lower grade materials or workmanship than that specified, even if the contractor offers a lower rate of charge than the bill rate for the specified material, unless the engineer agrees to this. Such a change is a variation requiring issue of a VO.
7. The RE must give immediate notice to the contractor when any defects in materials or workmanship are observed, because it may be very difficult to rectify a defective part of the work after it is completed. Hence inspections

of quality should take place as soon as an operation starts, or as soon as material to be used in the permanent works is delivered to site.

9.6 The resident engineer's duties with regard to safety

The safety regulations applying to construction in the UK are described in detail in Chapter 10. Under the CDM Regulations, there must be a health and safety plan drawn up by the employer's planning supervisor and extended by the 'principal contractor' to cover special or unusual aspects of the project. Primarily it is the responsibility of the contractor to comply with such a plan as needed for construction, and all safety regulations, as required by the ICE conditions of Contract (Clauses 8(3), 15(1) and 19(1)).

The RE will normally be appointed the engineer's Safety Manager on site. Hence he must ensure that his staff and any visitors he brings to site conform with all safety requirements and co-operate with the contractor. If the RE shows visitors round the site, he should advise the agent that he wishes to do so, and should see such visitors are accompanied when touring the site and have been informed of the safety rules. Normally all formal visits by outside bodies to view the project should be prior agreed with the contractor or his agent.

If the RE notices a failure by the contractor to comply with a statutory safety regulation or any site safety rule, he should inform the agent or contractor's Safety Supervisor and request compliance. The RE's request should be verbal in the first instance, since the failure might not have come to the notice of the agent or Safety Supervisor. If the correct safety measures are not adopted within a reasonably short time, a written note should be sent to the agent confirming the requirement. If the contractor still does not comply, the RE can instruct the contractor to comply, suspending the unsafe works if necessary or warning him that he proposes to call in the Health and Safety Inspector – but this action should not be adopted until all possible means of persuasion have failed. It would be impolitic of the RE to contact the Health and Safety Inspectorate without first warning the contractor. Also the RE must be sure of his grounds, and it must be borne in mind that a Health and Safety Executive (HSE) Inspector might not be available to visit the site immediately.

9.7 Relationship between the resident engineer and the contractor's agent

The RE must not be surprised to find that, on a new job, he is at first treated with considerable circumspection by the agent. He has to be, because one of the unknown factors the contractor has yet to discover which is of considerable importance to him, is what kind of RE will be in charge. The agent will need to go carefully at first so that he 'can get the measure' of the man who can daily affect the contract work. He will want to know what special matters are the concern of the RE and how he will wish to handle liaison between them. In like manner, the RE will be waiting to observe how competent the agent is and what degree of trust can be placed upon him, in order to find out what degree of supervisory control will have to be exercised.

The agent will want the RE to be fair, reasonable, and understanding. He will want clear decisive instructions from the RE, and prompt answers to his requests for information. He will want information and instructions about some work well before he starts on it; not after, or when he is part way through. He will object to an RE who is too keen on interfering in matters that should properly be handled by the contractor, or who makes contact with his subcontractors without the express permission of the contractor beforehand. He will expect all the RE's directions to be given only to him – except in cases justified by emergency.

This does not affect traditional practices adopted for contact between the RE's staff and the contractor's staff, such as when the RE's inspectors contact the agent's section foremen.

If the RE has any complaints, the agent will wish to be told about them personally. The RE should never make a complaint initially by letter. Such a letter will seem unfair to the agent, because a letter puts a complaint 'on record' before the agent has any chance to show the complaint is misplaced.

An especial nuisance to the agent is an RE who is too meticulous and rigid in his views – who thinks it necessary to measure up every cubic yard of concrete to the third decimal place; or who insists that every word in the specification must be exactly and rigidly complied with, irrespective of the need to apply such conditions in every case. To make reasonable judgements that are accepted as fair by both the contractor and the engineer, should be the principal aim of every RE.

9.8 Handling troubles

There will be times when troubles arise; such as when bad workmanship comes to light, or quite unsuitable methods are being used. It is the RE's duty to have the work rectified or the unsuitable methods stopped. This is easy to say, but not so easy to carry out in practice. The first requirement is that bad workmanship ought to be discovered at the earliest possible stage. The second is to be careful when having to point out defective work. Accusations are out of place; most defective work occurs through mishap, lapse of control, or because someone has been set to do a job beyond his competence. Nor should the RE start his complaint with some provocative remark which causes resentment and an inevitable row.

Instead, the RE should ask the agent to view the defective work with him, indicating that he has concerns about its acceptability. When they meet to view the defect together, the wise RE will say nothing, but will allow the agent to examine the matter for himself. One of two things will happen now: either the agent will make some admission of fault, or he will say, 'What's wrong with it then?' If the agent admits a fault there is no doubt that with careful handling all can be made well; but if the agent asks why the RE has his objection, the RE must tell him clearly why, what would have been acceptable, and what might be done in the circumstances. This opens the door to possible remedies, and further discussion may make it possible to discover the remedy which is cheapest to adopt.

However, if no acceptable remedy can be agreed upon, it is best to leave the matter for the time being, so that both parties have more time to think about the problem. Leaving a decision over for a day or two is often a way of discovering the best answer to a problem.

There will be other occasions when the RE is not at all sure what he should do, such as when he has to decide whether or not he will accept some method proposed by the agent. The agent has to think up ways of doing things that are cheapest, using the men and machines he has got. He may therefore propose methods which come as a surprise to the RE, who has been schooled to think in terms of using the 'right' way for each particular job. The old style general foreman was fertile in thinking up unusual methods of construction that saved him trouble, and not short of explanations as to why no possible harm could result.

The reasonable RE will not wish to deprive the agent of opportunities for benefiting from his own skill; on the other hand, he must not allow chances to be taken which might cause damage or early deterioration of the works. If, therefore, he permits the agent to proceed on his proposed method he would be within his rights to forewarn the agent that, if any harm does result, then the contractor must make the harm good at his own expense. If there is not time to discuss the matter with the engineer, the RE should discuss the problem with his own staff, because it is always useful to take others' opinions, and discussion can reveal important points that may have been missed.

9.9 More difficult cases of trouble

One of the most difficult things for the RE to tolerate is to stand by and see the agent make a mess of things. He cannot step in and tell the agent how to do his job, but he may see time wasted, unsuitable methods tried and abandoned, errors having to be rectified, and lack of control and proper planning. He may get to hear, in a roundabout way, of complaints from the agent's men about the way the job is run. He fears that all this indicates trouble in the future and does not know quite what to do about it.

It is necessary for the RE to wait until there is sufficient factual evidence to report to the engineer, such as poor progress and too much work having to be rejected; plus instances of obvious mistakes made by the agent when, for instance, some eccentric method of constructing some work has had to be aborted. But probably the most persuasive information likely to lead the contractor to withdraw an incompetent agent, is for the engineer to provide the contractor with the estimated value of the work done to date, compared with the contractor's probable expenditure. If the comparison shows an unacceptable loss to the contractor, the RE may be gratified to see how quickly a contractor can act to remove an incompetent agent. But even if the RE's estimate does not show a clear loss, the estimate will at least cause the contractor to examine what his agent is doing.

A difficult problem arises for the engineer if too many disputes seem to arise between his RE and the agent. The engineer has power to require the contractor to withdraw his agent under the ICE conditions (Clause 15), but he will be reluctant to use this power unless he has incontrovertible evidence the agent is solely at fault. If he suspects there is a clash of personalities on site, this can put both the contractor and engineer in a difficulty. There is danger that they may agree the problem can be resolved only by removing the agent or the RE. But the decision to remove either is then one of expediency and not necessarily justice, and it can damage the reputation of both the RE and agent.

To avoid such a situation arising, the RE must appreciate how his own conduct can affect the agent's reaction. One of the most certain ways of losing the agent's co-operation is to be 'continually reading the Specification at him' as if strict compliance with it applies to every situation however irrelevant. The agent will think that unreasonable – which it is. He will also regard lack of appreciation of his difficulties as unreasonable. When the agent faces difficulties and is in need of help, it is up to the RE to relax conditions that are not essential and to permit other ways round to the end result desired. An agent will never resent a call from the RE for especial care with some operation, or for strict compliance with the specification in matters of importance, such as for a top-class finish for those parts of the job which will remain permanently in view; but in return he will expect there will be occasions where the strict letter of the specification is unnecessary and will not be demanded by the RE if compliance presents real difficulty.

The contractor who continuously submits claims for extra payment, and will not withdraw them despite many being obviously invalid, presents a problem. The subject of claims themselves and how to deal with them is dealt with in detail in Chapter 17. The initial problem is how the RE is to deal with such a contractor. In the first instance, however many claims are submitted, it is essential the RE gives an immediate answer in writing to every such claim, registering any reasons for his non-acceptance of the claim. He must make his answers factual and courteous, and not let his letters show signs of irritation or complaint. The reason is that, if the dispute should go to arbitration, all correspondence relating to the claim must be put before the arbitrator. Thus if the RE's letters follow strict fact and are courteously worded, the more will his views impress the arbitrator.

The chief defence against a disputatious contractor who submits many claims is for the RE to maintain extensive records concerning every claim. The site diary, the weekly reports, the daily reports of the inspectors, copies of notes of instruction to the contractor and daywork sheets, and reports of tests, may all help to ensure that decisions on claims are supported by factual evidence. All must be filed in first class order. When meetings are held with the agent or contractor to discuss claims, minutes of the meeting should be drawn up by the RE, at latest by the day after the meeting, and submitted to the contractor for agreement. Inevitably the submission by the contractor of unreasonable claims is bound to cause a degree of coolness between the RE and the agent. But care must be taken not to let the situation decline into open hostility.

Under the ECC conditions the project manager 'may, having stated his reasons, instruct the contractor to remove an employee' (Clause 24.2). These reasons need to be soundly based and convincing if the project manager is not to sour his relationship with the contractor, especially if he requires removal of the contractor's agent, because the contractor has no redress if he thinks the reasons stated are inadequate.

The ECC conditions also try to deal with the problem of excessive claims by the introduction of 'early warning meetings' to deal with any matter which the contractor or project manager becomes aware could increase cost, cause delay or impair the performance of the works (Clause 16.1). 'Either the Project Manager or the Contractor may instruct the other to attend an early warning meeting. Each may instruct other people to attend if the other agrees' (Clause 16.2). Those who attend 'co-operate in making and considering proposals' to avoid or reduce the effect of matters raised, 'seeking solutions that will bring advantage to all those who will be affected', and deciding actions (Clause 16.3). Presumably the agreement of the third party to attend must be sought and the aim of the process is to resolve potential claim situations and disputes by agreement.

9.10 The resident engineer's staff

Except for the largest jobs the RE's staff on UK sites will be quite small. Two or three assistant engineers and two or three inspectors might be needed for a £25 million project in the UK; but much depends on the nature of the work. There is usually a considerable amount of work for assistant engineers to do during the first one-third period of a project, tailing off thereafter. On large jobs a measurement engineer or sometimes a quantity surveyor having experience of civil engineering work, may be needed to handle the checking of interim payments, dayworks sheets, etc. This can be important because, if the RE has only a couple of assistant engineers, he will not want to lose one on office work. A driver and suitable vehicles may be essential for getting about the site or carrying surveying equipment, taking concrete test cubes and soil samples for testing, etc. A chainman-cum-teaboy on the RE's staff must not be forgotten for his presence on even the smallest site can be a great asset. It is usual for the chainman, and the driver plus vehicle, to be provided by the contractor under the contract, and woe betide the drafter of the contract documents if he forgets to include provision of these in the specification.

Under the ICE conditions the engineer or RE must notify the contractor of 'the names, duties and scope of authority' of persons appointed to assist the RE in his duties (Clause 2(5)). This must include the names of inspectors as well as assistant engineers because the clause goes on to say that such assistants are not to have any authority to issue instructions save as 'may be necessary to enable them to carry out their duties and to secure the acceptance of materials and workmanship as being in accordance with the Contract.' This clearly implies they have power to accept or reject materials or workmanship. However, this power must be exercised with tact and understanding.

It is not sufficient to take the view that the RE and his staff are present solely to ensure the works conform to specification. To serve the engineer and employer properly they must assist the contractor make a good job of the construction. When unexpected conditions occur, assistance must be given to find a solution that is not only necessary for the quality of the permanent works but is also that which the contractor feels he can do satisfactorily. The queries the contractor raises must all be answered constructively and, when reasonable help is asked for, it should be given.

The *engineering assistants* should be kept informed of problems on the job, so that their actions can be intelligently directed. This helps to avoid mistaken or contradictory instructions being given to the contractor. Young engineers on site for the first time need to be forewarned of some of the troubles they can fall into. A young engineer may know it is injudicious of him to give the general foreman 'an instruction'. But he may not be aware that a question he innocently puts to a section foreman may (less innocently) be translated into 'a complaint' which, travelling rapidly upwards, brings an irate agent into the RE's office, asking 'What is this trouble your engineer is complaining about?' It all sounds rather difficult, but site life is rather a closed society which seldom resists the temptation to 'put a newcomer in his place' to start with. However, once relationships are established and statuses are recognized, such troubles blow over.

Status on site is tied to evident competence and the ability to give clear instructions courteously; it has nothing to do with rank or gender. Construction sites run by a good agent and a sensible RE can provide an outstandingly valuable and enjoyable experience to an engineer in his or her career.

The *inspectors* have to be mostly outside, watching the workmanship. Inspectors are usually older men, but this is no disadvantage because their practical experience is of value to the RE, and also an advantage when having to deal with the contractor's workers. Persuasion, tact, tolerance, care in observation, and the ability to give firm direction are required. Not everyone possesses these qualities, and it is not really the job for a young man who can find it irksome to watch the work of others he sometimes feels he could do better himself. On overseas sites an inspector plays a much more positive role, often having to teach and demonstrate to labourers how work should be done. A good inspector can be an asset to a contractor. One agent said 'A good inspector relieves me of some of my worries. When he passes something I know it should be all right.' One of the problems for the RE is how to get hold of a 'good' inspector. Usually it is best done by recommendation from an RE who has employed the inspector before. Some firms of consulting engineers keep good inspectors in continuous work, passing them from one job to another. If the RE hears of one such 'coming free' and can gain his services, he is lucky.

9.11 Gifts and hospitality

At Christmas – and other festive occasions – cheerful visitors may appear at the door of the RE's office, wishing him and his staff the season's greetings and perhaps extending some gifts. Politely but firmly, without giving offence, the good wishes may be accepted but not the gifts. No doubt the gift is innocently intended: the contractor or an 'approved' sub-contractor or supplier may be well pleased at the treatment he has received and wants to express his gratitude. But the engineer and all his staff occupy a position of trust in which all parties involved in a contract – the employer, the contractor, his suppliers and sub-contractors – expect to get fair treatment. To accept a gift from any of them, or any kind of pecuniary favour, might put in doubt the claim of the engineer and his staff to be impartial. It could be disastrous for the RE (and for the contractor) if, having to give evidence on some unhappy dispute arising under the contract, the RE has to admit under questioning he accepted gifts from a contractor.

The question of accepting hospitality is a different matter. It is uncivil to refuse all invitations of this kind; courtesy demands that on the right occasion hospitality should be accepted, and returned. The RE's common sense should tell him when it is right, such as when a triumph on a job is to be celebrated; when personnel depart from the job; or when troubles on the site need to be discussed in an 'off-the-job' atmosphere. As long as the giving and receiving of hospitality is conducted reasonably, these actions do much to promote friendly co-operation on the site for the benefit of the job.

10

Health and safety regulations

10.1 Legal framework

The law relating to health and safety has evolved on two fronts: by statute and through the common law. Until the Health and Safety at Work Etc. Act 1974 was enacted, there was no real framework to the law on health and safety in the UK. Previous legislation consisted of prescriptive rules which employers were required to follow. Lord Robens was instrumental in the development of health and safety law and his report of 1972 formed the basis for the Act. The Robens approach was to establish 'goals' rather than rules within a legal framework that required employers,, for instance, to ensure the health, safety and welfare of their employees at work. Subordinate legislation which followed the 1974 Act was designed on similar principles, however not all modern legislation is written in this way and some sets absolute standards to be followed, for example, use of explosives in demolition.

The 1974 Act is the principal UK legislation concerning health and safety and it is under this Act that the majority of health and safety regulations are empowered.

10.2 The Construction (Design and Management) Regulations 1994

The Construction (Design and Management) Regulations 1994 (SI 1994/3140) – known generally as '**the CDM Regulations**' – came into full effect on 1 January 1996. They implement EC Directive 92/57/EEC which set out minimum safety and health requirements at temporary or mobile construction sites. The regulations brought major changes to construction health and safety, explicitly bringing clients and designers into the process for the first time. The aim is to promote effective health and safety measures by placing certain duties on the client, designers and contractors involved in a project, and introducing a new role of **planning supervisor**. The Health and Safety Executive (HSE) administers and enforces the regulations. The following are the main requirements.

- The client has to appoint a planning supervisor and name the principal contractor and be reasonably satisfied that they, and also the designers, have adequate resources and competence to carry out their duties (Regulations 6, 8 and 9). He must provide the planning supervisor with any relevant information (Regulation 11) and ensure that information in any health and safety file delivered to him (see below) is kept available for inspection by persons needing to comply therewith (Regulation 12). The planning supervisor and principal contractor can be the same person, or the client himself can act as either or both (Regulation 6(6)).
- The planning supervisor has to notify the HSE of the intended project (Regulation 7) and ensure that a health and safety plan is prepared in respect of the project (Regulation 15(1)). He has to ensure that the designers pay adequate regard to health and safety matters (Regulation 14(a)(b)) and be in a position to give advice on the competence and adequacy of resources of designers and contractors (Regulation 14(c)). He ensures that a health and safety file is prepared for each structure (Regulation 14(d)), which includes relevant safety information and is kept up to date with any changes during construction (Regulation 14(e)). He must ensure that the file is delivered to the client on completion of the construction (Regulation 14(f)).
- The designers have to ensure that any design 'includes among the design considerations adequate regard to the need (i) to avoid foreseeable risks' to health and safety; (ii) to 'combat at source (such) risks'; (iii) to 'give priority to measures which will protect all persons carrying out construction work or cleaning work at any time and all persons who may be affected by the work of such persons' (Regulation 13(2)(a)). Designers must also ensure that the design includes 'adequate information about any aspect of the project or structure or materials (including articles or substances) which might affect the health or safety of any person' (Regulation 13(2)(b)). The foregoing requirements are to be met 'to the extent that it is reasonable to expect the designer to address them at the time the design is prepared and to the extent that it is otherwise reasonably practicable to do so' (Regulation 13(3)).
- The principal contractor is required to comply with the health and safety plan and augment its provisions as necessary during construction (Regulations 15(4) and 16(1)(e)). As principal contractor he has to co-ordinate the activities of all other contractors and sub-contractors on the site and see that they comply with the health and safety plan (Regulations 16 and 17). He must permit employees and self-employed persons to discuss and advise him on health or safety matters (Regulation 18). All contractors must comply with rules in the health and safety plan, and clients and self-employed persons must be informed of the contents of the plan or such part of it as is relevant to their work (Regulation 19).

Notifiable projects to which all the regulations apply are thos
struction 'will be longer than 30 days or will involve more than 5(
of construction work' (Regulation 2(4)). For 'domestic' work – exce
tion or housing estate developments (Regulations 3(8) and 3(3)) – and w
involving less than five construction workers at any one time (Regulation 3(2)), the design requirements (Regulation 13) only apply. 'Domestic' work is defined as work not carried out in connection with a client's trade, business or other undertaking (Regulation 2).

The initial intent of these regulations was to promote better safety standards in construction by integrating safety into project management. The production of a safety plan, specifically required to identify risks, should assist in this. But the designer's responsibilities for safety are difficult to assess, as it may be problematic to decide how far designs must be modified to reduce hazards if this involves substantial extra cost. Also both the designers and the planning supervisor have responsibility for ensuring that any design pays 'adequate regard' to the need for health and safety measures (Regulations 13(2) and 14(a)), so the question can arise as to who decides what measures are adequate.

10.3 The Health and Safety Plan required under CDM Regulations

The CDM Regulations require that a Health and Safety Plan be developed pre-tender and then continued and modified in the construction phase. The pre-tender plan is to help potential contractors understand the specific risks of a site and the work to be undertaken. It is drawn together from an assessment of the site and information from the designers. The aim is to target key issues and not to spell out the usual hazards of construction which should be apparent to any competent contractor. Too much detail may obscure vital matters. The pre-tender plan will be issued to tenderers but need not become a contract document; if it does there is a possibility that the plan could interfere with a contractor's freedom of choice of methods of construction and of dealing with hazards. However, if a client has specific safety rules, say on an existing works site, then these should be included in the contract documents as they are intended to become obligations on the contractor.

The pre-tender plan should include:

- Project description and details of client, planning supervisor and designers.
- Existing safety arrangements and rules, permits and emergency procedures.
- Safety hazards including: access, hazardous materials or structures, existing services and ground conditions.
- Health hazards including asbestos and contamination.
- Design assumptions and identified risks, co-ordination of future design changes.

- Safety goals for the project and arrangements for monitoring and review.
- Environmental restrictions and on-site risks.

The construction phase of the Health and Safety Plan will be drawn up by the Principal Contractor to account for risks noted above and to allow for methods of dealing with these risks and other risks arising on site and must include:

- Management structure and arrangements for monitoring health and safety matters.
- Liaison with other parties on site and with the workforce.
- Selection and control of other contractors and exchange of information.
- Site security, induction, training and welfare details.
- Production of risk assessment and method statements.
- Emergency and reporting procedures.

The above are of course only an indication of the requirements and each project will have its own particular needs. The HSE Approved Code of Practice (see Section 10.6) gives further guidance on these and other requirements of the CDM Regulations.

10.4 The Health and Safety File required under CDM Regulations

The purpose of the Health and Safety File is to provide clients, and those who may do work for them in future, with information on any residual risks remaining within the finished structure. This may be needed for cleaning and maintenance, future construction or alteration and eventual demolition. The planning supervisor has a duty to ensure that the file is prepared and passed to the client and the other participants have a duty to ensure that all relevant information is supplied accurately and promptly. As preparation of the file will be one of the last actions relating to the project it is important to decide early on who is responsible for producing it. The file should not contain records of past plans and risk assessments but should include:

- A brief description of the work and how any pre-existing hazards have been dealt with.
- Key structural principles, safe working loads and exclusions on types of loading.
- Any hazards with regard to materials used and for cleaning and maintenance.
- Information for removal of any plant or equipment and any related hazards.
- Location and marking of services including electric, gas, fire fighting systems.
- As-built information and drawings showing means of safe access and exit from all parts of the structure.

10.5 Training

As part of the drive towards safer designs and safer sites there has been a big increase in training in the years since the CDM Regulations were introduced. Staff and operatives at all levels are expected to have attained a suitable level of knowledge and competence for their role. For example a graduate engineer will be expected to know about the relevant legislation and regulations and understand the principles of risk control, while a senior engineer must be able to carry out risk assessments and apply control measures and advise his juniors.

The ICE has provided guidance on the health and safety competency levels expected for differing levels of seniority. The various duty holders under CDM have to be able to demonstrate both competence in their roles and that they have adequate resources available. This may involve demonstration of experience and track record from previous projects, personal levels of training, and top management commitment. Induction sessions are necessary for all those new to a design team or to a site, so as to explain to them any particular risks and the general safety practices in place. On site this could include use of personal equipment, site rules, permit to work areas, emergency procedures, welfare arrangements and specific training such as task based 'toolbox talks'. For those going to site for the first time the HSE publication *Health and Safety in Construction* (HSG 150) gives advice and guidance on safety in various work areas and has a useful checklist of common hazards.

There are many industry organizations providing safety training but a recent development has been the introduction of certification of staff under the Construction Skills Certification Scheme (CSCS). This requires the passing of a basic test of health and safety knowledge as well as specific tests of competence in any area of specialism such as for machine operators. Some parts of the industry are aiming for full registration of staff on sites by the end of 2003.

The ICE have launched a Health and Safety Register for engineers who wish to demonstrate a defined level of competency in the application of health and safety within the construction process.

10.6 Approved Code of Practice under CDM Regulations

To clarify requirements of the CDM Regulations the Health and Safety Commission (HSC) published an Approved Code of Practice and Guidance on the CDM Regulations in late 2001 called *'Managing Health and Safety in Construction'* (HSG 224) which came into force on 1 February 2002. Although it made no changes to the regulations, it set out in clearer terms the legal responsibilities imposed by CDM than a previous, earlier version of the Code.

The revised **Code** sets out **Directions** which are printed in bold type and have special legal status. If a person prosecuted for breach of health and safety law is proved not to have followed the relevant provisions of the code he or she will need to demonstrate compliance with the law by other means, or a court may find him or her at fault. The '**Guidance**' in HSG 224 (put in normal type) has a different legal status and is not compulsory. However, if its provisions are followed this should normally be enough to achieve compliance with the law. Both clients and designers need to appreciate the implications of the new Code of Practice because of its legal status, and its extension of their duties beyond those given in the CDM Regulations.

The new Code also emphasizes the need for the management of health and safety throughout the life of a project, and therefore HSG 224 includes key elements of the Management of Health and Safety at Work Regulations 1999.

10.7 The Management of Health and Safety at Work Regulations 1999

The Management of Health and Safety at Work Regulations – known as '**the Management Regulations**' – were first published in 1992 but were later revoked and replaced by the same titled regulations of 1999 (SI 1999/3242). They implement EC Directive 89/391/EEC (known as 'the Framework Directive') which was passed to encourage improvements in the health and safety of workers at work. Although the general provisions of the Directive were already covered by virtue of the 1974 Act the details of the European legislation needed to be enacted by means of regulations. Both the early 1992 version and the substituted 1999 version of the regulations have provided the backcloth for other regulations to be enacted. Five other 'daughter' EC Directives were introduced following the 'Framework Directive' and these have been implemented in further UK regulations. Together with the 1999 Management Regulations they are what have been called the 'Six Pack' Regulations. The further UK Statutory Instruments are:

- The Workplace (Health, Safety and Welfare) Regulations 1992.
- The Provision and Use of Work Equipment Regulations 1998.
- The Personal Protective Equipment at Work Regulations 2002.
- The Manual Handling Operations Regulations 1992.
- The Health and Safety (Display Screen Equipment) Regulations 1992.

These regulations are fundamental to modern principles of health and safety management and deal with **assessment of risk** and arrangements for **competence** in the measures needed to protect individuals and prevent accidents. While CDM is not part of the 'Six Pack', the Management Regulations will always apply in those circumstances where CDM does not apply. Schedule 1 of the Management Regulations further requires not only that risks should be avoided and combated at source, but also that those which are unavoidable should be evaluated. This is a new requirement and should be read in conjunction with CDM Regulation 13 on the duties of designers.

The principles of prevention set out in Schedule 1 of the Management Regulations are as follows:

(a) avoiding risks;
(b) evaluating risks which cannot be avoided;
(c) combating risks at source;
(d) adapting the work to the individual, especially as regards the design of workplaces, the choice of work equipment and the choice of working and production methods, with a view, in particular, to alleviating monotonous work and work at a predetermined work-rate and to reducing their effect on health;
(e) adapting to technical progress;
(f) replacing the dangerous by the non-dangerous or the less dangerous;
(g) developing a coherent overall prevention policy which covers technology, organization of work, working conditions, social relationships and the influence of factors relating to the working environment;
(h) giving collective protective measures priority over individual protective measures;
(i) giving appropriate instructions to employees.

10.8 Risk assessment

Risk assessment forms an integral part of the design function when a decision has to be taken between the risk to health and safety and other design considerations.

Risk is defined as the likelihood of potential harm from a hazard being realized. A hazard includes articles, substances, plant or machines, methods of work, the working environment and other aspects of work environment with the potential to cause harm.

The HSE have provided guidance on the execution of risk assessments in their 'Five Steps to Risk Assessment' leaflet. The document is aimed at workplace risks; however it is a useful model for engineers to follow during the design and construction phases of a project. Risk assessment is set out under the following steps:

1. Look for hazards.
2. Decide who might be harmed and how.
3. Evaluate the risks and decide whether the existing precautions are adequate or whether more should be done.
4. Record your findings.
5. Review your assessment and revise if necessary.

Engineers should also refer to the policy and procedures set out within their own company safety management system relating to the requirements of CDM and the Management Regulations and the undertaking of risk assessments.

It is important to remember that having identified the hazards and those at risk (step 1 and 2) it is necessary to assess the level of risk (step 3) in order to decide on the order of significance and the preventive action needed. The approach here may be qualitative, based on subjective judgements, or quantitative using numerical estimates of risk based on probability and severity derived from empirical data. In certain industries, particularly where high risks are involved (e.g. chemical and nuclear industries), the quantitative approach is necessary in order to comply with legal or licensing requirements and this requires a more complex risk assessment methodology driven by statute.

Many people think that the risk assessment process ends when the risks have been assessed or ranked but this is incorrect. Giving risk a 'number' or rank is only the first phase in the management of risk and appropriate measures then need to be identified in order to eliminate or reduce the risks to as low a level as is reasonably practicable.

Reasonably practicable

CDM Regulation 13(3) requires the design to include matters '... to the extent that it is reasonable to expect the designer to address them at the time the design is prepared and to the extent that it is otherwise reasonably practicable to do so.'

The term 'reasonable' or 'reasonably practicable' is used in many of the post 1974 Act legislation and its meaning can be obtained by reference to common law judgements:

> '*Reasonably practicable*' (implies) that a computation must be made ... in which the quantum of risk is placed on one scale and the sacrifice ... for averting the risk (whether in money, time or trouble) is placed on the other ... .
>
> *Asquith AF; Edwards v National Coal Board* (1949)

The HSE provide guidance sheets and many other publications to assist the designer with ensuring the requirements of Regulation 13 are satisfied. CIRIA[1] report R166 is also recognized as a valuable publication intended to assist designers of construction projects to produce schemes that are safer to build and maintain. It provides essential guidance on the identification of hazards in relation to the health and safety of construction workers and those affected by construction work. It shows ways in which hazards can be avoided, reduced or controlled, together with options designers may be able to employ to comply with the CDM Regulations.

[1]The Construction Industry Research and Information Association, London.

10.9 The Construction (Health, Safety and Welfare) Regulations 1996

The 1996 Construction Regulations as titled above (SI 1996/1592) completed the implementation into UK law of EC Directive 92/57/EEC referred to in Section 10.2 above, and replaced many previous regulations. They address specifically the following:

- safe places of work;
- prevention of falls;
- falls through fragile material;
- falling materials or objects;
- the stability of structures;
- demolition and dismantling work;
- explosives;
- excavations;
- cofferdams and caissons;
- prevention of drowning;
- traffic routes, doors and gates;
- plant and equipment;
- vehicles;
- the prevention of fire and flooding;
- emergency routes and exits;
- emergency procedures;
- fire detection and fire fighting;
- welfare facilities;
- fresh air;
- temperature and weather protection;
- lighting;
- site tidiness, good order, site demarcation.

They apply to all construction works and also deal with the practicalities of work on site rather than the management issues covered by CDM. Guidance on the 1996 Regulations and other legislation is published in Health and Safety in Construction (HSG 150) which is recommended reading for engineers before they visit site for the first time.

10.10 Other major regulations

In addition to the legislation covered earlier in this chapter the following are worthy of note.

Provision and Use of Work Equipment 1998 Regulations (PUWER) generally require that equipment provided for use at work is suitable, safe for use, maintained in a safe condition, and in certain applications inspected. Personnel using work equipment should have adequate information, instruction and training. Suitable safety measures, for example, protective devices, markings and warnings should also be in place.

Lifting Operations and Lifting Equipment Regulations 1998 (LOLER) require that any lifting equipment used at work for lifting or lowering loads is:

- strong and stable enough for the particular use and marked to indicate safe working loads;
- positioned and installed to minimize any risks;
- used safely, ensuring the work is planned, organized and performed by competent people;
- subject to ongoing thorough examination and, where appropriate, inspection by competent people.

Workplace (Health, Safety and Welfare) Regulations 1992 apply to most but not all workplaces, some workplaces such as mines, quarries, construction and temporary mobile work-sites, and offshore installations are covered by separate legislation. The regulations aim to ensure that workplaces meet the basic health, safety and welfare needs of all the members of the workforce including people with disabilities and need to be considered during the design stage of projects. Health issues covered by the regulations include:

- adequate ventilation;
- temperature in indoor workplaces (thermal comfort);
- lighting;
- cleanliness and waste materials ;
- room dimensions and space;
- work stations and seating.

New Roads and Street Works Act 1991 covers works undertaken within the highway such as inspection, placement and maintenance of pipes, cables, sewers, drains, etc. which are laid in the carriageway or footway. It does not include road construction or maintenance which is covered by the Highways Act.

Control of Substances Hazardous to Health Regulations 2002 (COSHH) revoke and re-enact, with modifications, the Control of Substances Hazardous to Health Regulations 1999. They include changes to implement the requirements of the Chemical Agents Directive. COSHH applies to those substances classified as very toxic, toxic, harmful, corrosive or irritant under the Chemicals (Hazard Information and Packaging) Regulations. The regulations require a risk assessment to be undertaken of health risks created by work involving substances hazardous to health.

Control of Lead at Work Regulations 2002 revoke and re-enact, with minor modifications, the Control of Lead at Work Regulations 1998. The regulations also include changes required to fully implement the Chemical Agents Directive. Generally the regulations place duties on employers to provide greater protection to workers by reducing their exposure to lead.

Control of Asbestos at Work Regulations 2002 have been issued and came into force on 21 November 2002 except for Regulation 4 (21 May 2004) and Regulation 20 (21 November 2004). The regulations revoke, consolidate and re-enact with modifications of the Control of Asbestos at Work Regulations 1987 and

- introduce a new regulation to manage asbestos in non-domestic premises;
- incorporate the requirements of the Chemical Agents Directive;
- introduce a requirement for accreditation of laboratories that analyse materials to identify asbestos.

Manual Handling Operations Regulations 1992 outline how to address risks to health and safety of employees required to carry out manual handling in the course of their employment. HSE guidance refers to a maximum weight limit of 25 kg to be lifted and 20 kg for repetitive work such as laying blocks.

Health and Safety (Display Screen Equipment) Regulations 1992 places duties on employers relating to employees who regularly use computers, etc.

Electricity at Work Regulations 1989 require precautions to be taken against the risk of death or injury from electricity during work at or near electrical systems (electrical installations and equipment).

The Personal Protective Equipment at Work Regulations 2002 came into force on 15 May 2002. They repeal and replace the Personal Protective Equipment (EC Directive) Regulations 1992. The 2002 Regulations apply to anyone who manufactures, imports, supplies or distributes personal protective equipment (PPE). The regulations set out the basic health and safety requirements for different types and classes of PPE, including design principles and information that manufacturers must supply.

The Dangerous Substances and Explosive Atmospheres Regulations 2002 (DSEAR) implement safety aspects of the Chemical Agents Directive (CAD) and the Explosive Atmospheres Directive (ATEX 137). DSEAR covers safety aspects such as fires and explosions arising from dangerous substances and places requirements on employers and the self-employed to

- carry out a risk assessment of activities involving dangerous substances;
- eliminate, reduce and control identified risks;
- classify hazardous places into zones based on the likelihood of an explosive atmosphere being present;
- make arrangements for dealing with accidents, incidents and emergencies.

The Building (Amendment) (No. 2) Regulations 2002 have been issued and come into force on 1 July 2003. Regulation 2(8) and Schedule 1 come into force on 1 March 2003 and Regulation 2(7) on 1 January 2004. Primarily the following now applies:

- a new regulation requiring the carrying out of sound insulation testing in certain circumstances;

- a new Part B of Schedule 1 (Fire Safety) which replaces the existing Part B and new requirements on internal fire spread (linings) in Paragraph B2(1) allow for testing to European standards.

Control of Major Accident Hazards Regulations 1999 (COMAH) Their main aim is to prevent and mitigate the effects of those major accidents involving dangerous substances, such as chlorine, liquefied petroleum gas, explosives and arsenic pentoxide which can cause serious damage/harm to people and/or the environment. The COMAH Regulations treat risks to the environment as seriously as those to people.

The Confined Spaces Regulations 1997 apply where the assessment identifies risks of serious injury from work in confined spaces. These regulations contain the following key duties:

- avoid entry to confined spaces, for example, by doing the work from the outside;
- if entry to a confined space is unavoidable, follow a safe system of work;
- put in place adequate emergency arrangements before the work starts.

Work in Compressed Air Regulations 1996 provide a framework for the management of health and safety risks by those undertaking tunnelling and other construction work in compressed air. They address such issues as

- safe systems of work; medical surveillance;
- compression and decompression procedures (including HSE approval of procedures);
- medical treatment;
- emergency procedures;
- fire precautions;
- provision of information, instruction and training;
- maintenance of health and exposure records.

Many of the duties are placed upon compressed air contractors to reflect the practical operation of the industry and in recognition of the fact that the contractor in charge of the compressed air operations is best placed to manage and control the health and safety risks of such work.

Diving at Work Regulations 1997 seek to control the hazards associated with diving at work. They apply to all commercial diving in Britain. Practical guidance on how to comply with these regulations is contained in the HSC Approved Codes of Practice (L 104).

Reporting of Injuries Diseases and Dangerous Occurrences Regulations 1995 (RIDDOR) covers the duties of reporting of serious and fatal accidents, diseases and dangerous occurrences to the HSE by the employer or the controller of a site.

The Health and Safety (First-Aid) Regulations 1981 require employers to provide adequate and appropriate equipment, facilities and personnel to enable first aid to be given to employees if they are injured or become ill at work. These regulations apply to all workplaces including those with five or fewer employees and to the self-employed.

The Noise at Work Regulations 1989 and various other regulations apply to noise or include specific provisions on it, including: the Management of Health and Safety at Work Regulations 1999; the Provision and Use of Work Equipment Regulations 1998 and the Supply of Machinery (Safety) Regulations 1992.

A new Directive on the minimum health and safety requirements regarding exposure of workers to the risks arising from physical agents (noise), which will repeal Directive 86/188/EEC, was adopted in early December 2002. On 15 February 2003 the Directive came into force and the UK has three years to implement the requirements of it. The main aspects relate to the reduction in the trigger levels, that is, from 85 dB(A) to 80 dB(A) for making protection available and from 90 dB(A) to 85 dB(A) for hearing protection to be worn.

Pressure Equipment Regulations 1999 cover a wide range of equipment such as, reaction vessels, pressurized storage containers, heat exchangers, shell and water tube boilers, industrial pipework, safety devices and pressure accessories.

The Supply of Machinery (Safety) Regulations 1992 require all UK manufacturers and suppliers of new machinery to make sure that the machinery which they supply is safe. They also require manufacturers to make sure that:

- machinery meets relevant essential health and safety requirements (these are listed in detail in the regulations), which include the provision of sufficient instructions;
- a technical file for the machinery has been drawn up;
- there is a 'declaration of conformity' (or in some cases a 'declaration of incorporation') for the machinery;
- there is 'CE' marking affixed to the machinery (unless it comes with a declaration of incorporation).

Publications

Websites for locating Health and Safety Regulations and publications are:

legislation.hmso.gov.uk
hse.gov.uk
hsebooks.co.uk
rospa.co.uk

The address for Health and Safety Executive publications is:

HSE Books,
PO Box 1999, Sudbury,
Suffolk CO10 2WA, UK.

11

Starting the construction work

11.1 Pre-commencement meeting and start-up arrangements

Once an award of contract has been made, a meeting is necessary with the selected contractor to make preparations for starting the contract. Such a pre-commencement meeting will be attended by the employer or his key staff concerned, the engineer and his proposed resident engineer, and the contractor's manager and agent. This is to effect introductions and exchange information about the principal initial matters concerning each party. Items which may need to be covered by this meeting are as follows:

- Exchange of addresses, telephone numbers, etc. and establishing agreed lines of communication.
- Clarifying the resident engineer's delegated powers, and advising the contractor of his proposed staffing and supervisory arrangements.
- The contractor's report on the agent's experience in the type of work involved, in order that his appointment can be approved by the engineer.
- Any particular needs for temporary works designs or special methods of construction proposed.
- Arrangements for provision of sets of contract documents to the contractor and indication of any further drawings that will be supplied (e.g. bar schedules).
- Progress by the contractor in obtaining bonds and insurance (this is especially important where early access to site is expected since this may not be permitted until bond and insurance are secured).
- Proposed date for commencement, which, if not agreed, will be set by the engineer after taking the views of the contractor concerning his readiness to mobilize, and of the employer concerning the readiness of the site for occupation.

- The programme for construction which the contractor is to produce within 21 days of award of contract, and the consequent needs of the contractor in respect of further information and drawings to prevent delay.
- The contractor's health and safety plan and how this will work in conjunction with the employer's and engineer's responsibilities for safety.
- Provisions for access to the site the employer may require for his own staff.

Other matters which it might be important to consider include the siting of the resident engineer's offices, the services the contractor is to provide him with, and the layout of forms for monthly statements. This may not cover all the matters that may have to be discussed, and a further meeting may be needed to consider certain matters in more detail.

The success of this meeting in establishing good working relationships can make an important contribution in setting the tone for subsequent co-operation.

11.2 The contractor's initial work

The contractor's agent will probably come to site with a small nucleus of permanent employees, and his main aim will be to get started on the actual work of construction as soon as possible. He will have to visit the local employment office or employment agencies to make arrangements for taking men on site. The agent will find it necessary to have some clerical assistance on site from the start; for preference his site co-ordinator and office manager will accompany him and will start getting to site a wide variety of equipment, machinery and materials. Some of this will be sent out from the plant and equipment depot of the contractor's head office, but a large amount of supplementary equipment may be required from local sources. Consumables will be required: a term meaning all those things – picks, shovels, tools, fuel, timber, office stationery, protective clothing, lighting equipment, temporary fencing, furniture, canteen equipment and a legion of other items – which are not plant nor large items of re-usable equipment. A visit to the local bank manager may be necessary to make arrangements for withdrawing money for cash payments.

Plant may have to be hired for the work of digging trenches to lay water supply and drainage, and a dozer for site clearance. A gang of men may have to be set fencing off the site area, another gang on making foundations for huts, and a third gang on access road requirements. A site engineer will quickly be necessary for the setting out of levels and for producing sketches so as to direct the foremen and gangers what to do.

The agent will need to start arranging for delivery to site some of the materials required for early incorporation in the works, particularly the aggregates proposed for concrete, or samples of ready-mix from local suppliers. Such samples will have to be made into cubes and tested. This sampling and testing can take a long time, so must be started early if good quality concrete is required early on the job. The agent may visit – probably with the resident engineer – local suppliers of ready-mix concrete to observe their quality control, and to discuss rates of supply and qualities of concrete required.

It depends on the location of the site, the standing of the agent, and the policy of the contractor, how far materials for use in the works are ordered by the agent or by the contractor's head office. The supply of major materials for which head office already possess quotations would probably be ordered by head office. But the agent may need to order some supplies locally. He will probably seek to avoid entering long-term supply agreements with a new supplier for materials until he is confident the supplier will not default on deliveries or on quality of materials supplied.

It sometimes requires the combined efforts of the agent and resident engineer to get early installation of services such as telephones, power lines, sewer connections and water supply. On overseas projects the procurement of local materials, and the checking and steering of imported materials through customs often forms a major departmental function within the contractor's local organization.

11.3 The resident engineer's work

The engineer who finds himself newly appointed to take up the position of resident engineer and who has previously had little experience outside may well feel somewhat alarmed at the prospect before him. He has no doubt been told he 'will manage all right', but this seems small comfort as he thinks of all the things he does not know about the job and all the unknown questions likely to arise. He may also feel uneasy at the prospect of having to tell everyone what to do (instead of deciding action within a team) and may wonder how he is likely to match up to the contractor's agent who appears a tough and forceful character considerably older than himself. However it is unlikely that problems of any engineering magnitude will be immediately encountered, for there are many organizational details to deal with first.

Work before going to site

The resident engineer should have spent some time before he goes to site examining the contract drawings and specifications, and there should have been an opportunity for him to have conversations with the designers. He should get to know how the job has been designed, so that he is able to make intelligent suggestions if the conditions revealed during the course of construction differ from those expected. He should make a file of all information which is basic to the job, such as:

- soil test data on which the design has been based;
- geological information;
- levels and benchmarks used;

- rainfall and runoff data;
- details of special materials or equipment to be incorporated in the job;
- addresses of authorities and personnel who have been written to about the job, such as the local planning authority, the district road engineer, the local building surveyor, the employer and his directors or councillors and staff;
- a brief history of how the job came about and the dates and references of major decisions.
- the pre-tender Health and Safety Plan.

The compilation of this file of data can act as a check on the situation to date. A separate file of matters still outstanding is advisable. Once the resident engineer is appointed, everyone previously connected with the job will expect him to take responsibility for seeing that all site matters are done in due time. Thus the programme of construction to be agreed between the contractor and the engineer will be one of the documents carefully studied by the resident engineer, so that he can check it in detail for its consequences.

The resident engineer must make sure he has with him a final copy of the contract documents as awarded.

The site office

The resident engineer may have a choice as to where his office should be placed. If so, it should be placed so that, from it, the main traffic in and out of the site can be observed. It is a mistake to choose a situation which overlooks the job but which does not have a view of the main entrance. Little worthwhile of the job can be seen from a distance, whereas even a distant view of the entrance will enable the engineer or his staff to notice a number of happenings – the delivery of materials, plant going off site, and when callers are about to descend (especially the employer).

The office itself can range from a simple hut to a veritable barracks, according to the size of the job. On a moderate sized job where the resident engineer has two or three engineers to assist him he will need

- his own room;
- a drawing office for engineers;
- a secretary's office and filing room;
- a washroom and toilet;
- a small kitchen area where hot drinks can be made;
- a room where wet clothes can be stripped off and hung up to dry;
- a small store room for surveying and other equipment.

On many civil engineering jobs a soils and materials testing laboratory is necessary, and this is more conveniently placed near the resident engineer's offices than elsewhere. Outside the entrances to offices an essential item of equipment is a boot scraper, preferably with a small area of concrete with a hose-pipe water supply for cleaning gumboots.

11.4 Early matters to discuss with the agent

Items to be discussed will almost certainly concern the laying on of services to the job – telephone, water supply, electric power and drainage. Even with use of mobile telephones a land line is required as quickly as possible and the telephone authority may need assistance in getting permission to run lines across private properties. The agent may ask the resident engineer to approve proposals for hard standing for cars and the routing of access roads.

The question of *drainage and sanitation* may prove difficult to solve. The resident engineer has to watch that the contractual requirement to provide a 'small sewage treatment works' does not get whittled down to no more than a tank and a soakaway, or a tank and an overflow to a near-by ditch or river. The sewage works must be large enough to treat all the sewage from the maximum number of persons who will be employed on site plus an addition for visitors. If they are later found inadequate, it may prove difficult to get action if the contractor feels that, given a few more weeks, the number of men on the job will decline and the problem will solve itself.

The question of *waste oil disposal* from plant is a thorny one, and should be brought to the agent's notice. Discharge of used lubricating oil or waste diesel oil to public sewers is usually forbidden; to discharge it through the site sewage works will probably ruin their proper functioning. The discharge of even small quantities to a watercourse will almost certainly be detected by the Environment Agency who will demand immediate rectification and the contractor may be liable to a penalty and payment of compensation if damage has resulted. The waste oil should be led to a pit and disposed of by tanker as the local sewerage authority advises.

The resident engineer will need to know what part of the job the agent intends to tackle first, so that he can check any necessary setting out that must precede it. The agent will need to know what are the local *benchmarks* which have been used for the original survey of the area. If these are some distance away, they may both agree that their staff should jointly arrange for a convenient benchmark and base line to be set out near the job.

The next topic may be the *programme* as a whole, and this is the first of many discussions that will occur on that subject. Sometimes the agent wants more information from the resident engineer so that he can continue making his detailed plans, or he may have perceived some problem ahead which he thinks might be avoided if the engineer would sanction some action not exactly in line with contract requirements. The resident engineer had best give only a guarded opinion if this is his first acquaintance with such a proposition.

The resident engineer should be wary of discussing, too early, design matters or alteration of the contract requirements, because he may find out later that there are good reasons for the design requirements being as shown in the contract documents. Too early a desire to assent to some proposal by the contractor can lead to later trouble, when the assent has to be withdrawn as a result of increased understanding of the job.

11.5 Some early tasks for the resident engineer

At the end of the first week the resident engineer will no doubt find that he already has a number of tasks to do. If *bulk excavation* is about to start it will be essential to take levels of the natural ground over the site where the excavation is to take place, if these levels are not already available in sufficient detail. This is urgent work, for there will be no chance later of finding what the natural ground levels were and the calculations for quantities of excavation would then be largely 'intelligent guesswork' or agreement will have to be sought on bill quantities, which may differ from the true quantity excavated. If the contractor has taken his own levels over the site and the resident engineer has let pass the opportunity of checking them, he will be in no position to argue against the contractor's figure for the excavation.

It may not be sufficient to rely on ground levels shown on the contract drawings because these may be based on interpolation of published contoured maps of the area. Where such contours originate from aerial photography they can be a metre in error because they may reflect the top of vegetation rather than the soil level.

Another early task is to carry out and agree with the contractor the *state of existing buildings* which might be affected by construction of the contract works, and the state of *approach roads* to the site. This is essential so that any claims for compensation for damage can be decided properly. In this survey sets of photographs of existing cracks or damage, as well as general views, form an important part.

The resident engineer must see that all *productive top-soil* is stripped and stacked separately for later re-use. All amounts of soil should be so stacked, even that taken off areas for the site offices, since there is often a lack of soil at the end of the job. The question of disposal of excavated material will have to be considered. In many countries there is now increasing control over what materials may be disposed of to landfill and these need to be borne in mind when agreeing with the contractor what to do with unwanted excavated materials.

The next task the resident engineer may need to do, if he has not done it already, is to check the *delivery times* for any equipment or materials to be supplied under other contracts or by the employer, such as the supply of pipes and valves. On overseas jobs there may be many separate contracts for the supply of materials. All these separate supply contracts have to be checked in detail to ensure that nothing has been missed.

11.6 Meeting the employer

Shortly after his arrival on site the resident engineer should see if the employer wishes to meet him and will set aside a morning or afternoon for going over the site and discussing the project with him. It frequently happens that the employer or his representative does wish to keep contact with the job, but any observations of consequence the employer makes which might require some action, should be passed through to the engineer so that he may give the necessary directions.

The employer may, for instance, be hoping that certain sections of the work can be completed and made use of by him before completion of the job as a whole; or he may want certain sections left for the time being because he may be having ideas of altering his requirements. Both these matters impinge directly upon the contractor's programme and could change the cost of the job. Therefore, they have to be looked into by the engineer. Of course, if the employer is merely wanting to 'sound out' what is possible and how much it might cost, the resident engineer should give him a reasonable answer but make clear that the engineer must be involved before any decision is reached.

One other matter the employer may wish to raise is the traffic or noise created by the contractor about which the employer has already received complaints. The resident engineer may have to consider what reasonable requests he could put to the contractor which would reduce these complaints.

As some structures begin to take shape an employer can be expected to take more interest, and he may start making requests for minor additions once he or his operational staff see what the structure looks like. The resident engineer and engineer must expect this and, if the contract has been wisely drawn up, it will allow for some flexibility of requirements in the later trades. Many finishes, and particularly colour schemes, are best left for the employer to choose. There is no point in an employer paying large sums for a project and not having some choice as to its final appearance.

The resident engineer will endeavour to meet reasonable requests by the employer; but if some apparently extravagant extra is asked for, he should be wary. He must remember that where the 'employer' is a public authority, the members of that authority might not necessarily agree with any proposed extra expense suggested by an officer acting on their behalf.

Where construction of the works is likely to be regarded as a nuisance by nearby residents, the employer may have already set up a liaison committee with representatives of the residents to smooth out possible difficulties. The resident engineer should find out what has been agreed so that he can direct the contractor accordingly. He may also need to attend meetings with local community representatives to report on progress and future activities and to try to find ways of minimizing any nuisance caused by construction of the works.

11.7 Setting up the clerical work

It will be necessary to set up a system for the handling of correspondence, measurement of quantities and checking of contractor's interim payment applications, and for log sheets of all technical data. Details of what is required are set out in Chapter 13. A word processor and a copying machine will be an essential part of the equipment required. To check the contractor's interim payment application a print-out calculator, as used by accountants, is useful. This prints out the figures added so that checking for arithmetic errors is made easier.

Petty cash must not be forgotten, and the recipe is 'enough but no more' because of the risk of break-ins and theft. Petty cash never seems to balance (whatever accountants say) when the sum total of what it should be comprises a miscellany of stamps, a variety of small change, some crumpled notes, a bunch of folded receipts, and list of expenditure in practically everybody's handwriting. A deficit one week can become a surplus the next, and vice versa. If a deficit persists there is probably no criminal reason for it save human forgetfulness and should the resident engineer make it up from his own pocket he will perceive the wisdom of not having too much petty cash.

12

Site surveys, investigations and layout

12.1 Responsibility

The responsibility for setting out the works usually lies entirely with the contractor who will work to the dimensions shown on drawings and from levels or reference points given on the contract drawings or notified to him. Benchmarks for levels may be national marks, such as Ordinance Datum in the UK, or they may be special marks set up locally as used by the original surveyors for design of the works. Where the absolute level of the works is not critical in relation to other structures a local level mark may be chosen, preferably given high enough value to avoid negative levels in the deepest expected excavations. From the given benchmarks, levels should be brought to convenient benchmarks on the site itself in positions which are unlikely to be disturbed. Existing structures, or cast concrete blocks should be used with the exact point marked clearly and precisely levelled in.

Location of the works in plan is provided by reference to national grid points or to a local reference line. A series of fixed points must be provided around the site to allow accurate setting out relative to the reference. If these are to last any time they will need to be set in concrete or on existing structures and marked and protected. The exact position of these survey points can be established by triangulation where suitable Ordinance or other points exist, or by traverse using electromagnetic distance measurement (EDM) or total station equipment (see Section 12.3). It must be emphasized that establishing the site level and survey points accurately is essential. Many contractors will have professional survey staff to carry this out, but otherwise the work may be subcontracted to a specialist firm of surveyors.

Once the basic stations are established the contractor's engineers will set out detailed grid lines, levels and sight rails to allow construction to proceed. The resident engineer's staff should check critical lines and levels, although they do not need to replicate all the setting out work. Although the contractor remains responsible for setting out errors both he and the engineer have a duty to see that the works are properly constructed to line and level.

12.2 Levelling

For most site survey work tilting or automatic levels provide for quick and accurate results. As with all instruments, they should be checked regularly for accuracy and returned to the manufacturer for overhaul and re-setting at intervals. The levelling staff must be kept clean and its markings clear to reduce reading errors and it is worth spending time to ensure the chainman is clear as to where the foot of the staff should be placed and that it is held vertically. The tripod must be in good condition, and set up so as to avoid movement during operations. Once fixed to the tripod the instrument must be levelled using the centre bubble to avoid excessive tilt of the telescope. For the tilting level the split view of the tubular level bubble must be adjusted to coincide for each reading while the compensation mechanism in the automatic level ensures that the sighting is in the horizontal plane. On-site checks on accuracy can be made using the two-peg test.

Even with a good instrument and set up, care must be taken to ensure the levelling staff is correctly extended and that sighting lengths are not so great that readings are indistinct. Good visibility is necessary and may be reduced by heat haze or vibration in the wind. Accuracy of reading and of booking the readings are also sources of error. Some errors can be reduced by keeping sight lengths approximately the same and it is a good practice to close the run of levels back onto a known benchmark.

12.3 Plane surveying

In the past, most survey work depended on triangulation from known fixed points using a theodolite and this may still be a suitable method for smaller sites. Again it is necessary to ensure the instrument is in good condition and that its base is truly horizontal. Readings taken on both faces of the instrument may reduce residual errors. Setting out by taping along a line given by the theodolite may also still be the clearest way of providing centre lines or points, particularly for regular structure layouts such as building columns. The appropriate time for this is when blinding concrete has been placed to column and wall foundations. The base line, which is either the centre line of the building, or a line parallel to it but clear of the building, should have been set out previously by end pegs sited well clear of the work. It is usual to work from co-ordinates along this base line from some fixed zero point, and measuring right angle distances out from them. In this way lines of walls and column centres can be marked on the blinding concrete.

Distances may be measured by steel or fibreglass tape pulled horizontally, so it is a great convenience if the site is level. If not a plumb bob has to be used to transfer distances. Distance co-ordinates along the base line from the zero peg are set out, using the steel tape and marking a pencil line across the peg. The theodolite is set out over the pencil line, and its position is adjusted laterally so that it transits accurately on the two outermost base line marks. The plumb bob on the theodolite gives the mark for the co-ordinate point, a round headed nail being inserted on this point. Distances at right angles to the base line are then set out with theodolite and steel tape. The advantage of this method is that the theodolite can sight down into column bases which are usually set deeper than the general formation level. For the assistance of bricklayers and formwork carpenters, sight boards can be provided, with the cross-arm fixed at a given level above formation level and with saw cuts exactly on the lines of sight to be used. A builder's line can then be fixed through such saw cuts. An alternative to the foregoing is to set out two base lines at right angles to each other and use theodolite right angle settings from these to give centres for such column bases, etc.

The introduction of EDM equipment has, however, meant that accurate distance and angle measurements can now be made from a single point set up. The instruments work by measuring the time of a wave in travelling from the transmitter to a reflector and back. Readings may be automatically repeated to improve accuracy. Built-in or add-on equipment allows for automatic data logging, reduction of distances to horizontal and vertical components and for downloading to a computer. Accuracy over short distances is good. Over longer distances corrections may need to be made for atmospheric conditions which vary from the manufacturers' setting. The improved accuracy available has meant that setting out on site or general survey work is often done by some form of traversing. By this method the position of two known points is extended by noting the angle to a third point and its distance from the instrument set up over one of the points. Extended traverses should be closed onto another known point to check for errors.

Even with EDM equipment, setting out of regular structures is probably best done using a marked baseline as described above. The equipment also has major advantages in ground surveying since the location and elevation of any point in the area to be surveyed can usually be determined directly from just one or two positions of the instrument. Data from the instrument can then be downloaded into a computer and with the use of appropriate software, contoured plans of the area can be produced for design or for earthworks measurement purposes. A certain amount of planning is necessary to produce the best results by ensuring a regular grid of locations is used for targeting and that any individual feature, such as sharp changes in slope are picked up. As an alternative ranging poles can be used to set out a rough grid and readings at say 20 m intervals between these should give sufficient coverage for accurate plotting.

12.4 Setting out verticality, tunnels and pipelines

As a building rises the vertical alignment must also be controlled. This can be done by extending building centre lines at right angles to each other out to fixed points clear of the structure. These lines can then be projected up the building and marked, allowing accurate measurements from these marks at each floor. Alternatively an optical plumb can be used to project a fixed point up through openings in the floors of the building so as to provide a set of reference points at each level.

The standard of setting out for tunnels must be high using carefully calibrated equipment, precise application and double checking everything. An accurate tunnel baseline is first set out on the surface using the methods described above. Transference of this below ground can be done by direct sighting down a shaft if the shaft is sufficiently large to allow this without distortion of sight-lines on the theodolite. With smaller shafts, plumbing down may be used. A frame is needed either side of the shaft to hold the top ends of the plumb-lines and to allow adjustment to bring them exactly on the baseline. The plumb-line used should be of stainless steel wire, straight and unkinked, and the bob of a special type is held in a bath of oil to damp out any motion. By this means the tunnel line is reproduced at the bottom of the shaft and can be rechecked as the tunnel proceeds.

Many tunnels are nowadays controlled by lasers, the laser gun being set up on a known line parallel to the centre line for the tunnel and aimed at a target. Where a tunnelling machine is used, the operator can adjust the direction of movement of the machine to keep it on target so that the tunnel is driven in the right direction. For other methods of tunnelling, target marks can be set on the soffit of rings, the tunnel direction being kept on line by adjusting the excavation and packing out any tunnel rings to keep on the proper line.

Lasers are also used in many other situations, usually for controlling construction rather than for original setting out since their accuracy for this may not be good enough. The laser beam gives a straight line at whatever slope or level is required, and so can be used for aligning forms for road pavements or even laying large pipes to a given gradient. For the latter, the laser is positioned at the start of a line of pipes and focused on the required base line. As each new pipe is fitted into the pipeline a target is placed in the invert of the open end of the pipe, using a spirit level to find the bottom point, and the pipe is adjusted in line and level until the target falls on the laser beam. Bedding and surround to the pipe are then placed to fix the pipe in position.

Rotating lasers are also widely used and once set up give a constant reference plane at a known level. Use of a staff fitted with a reflector allows spot levels to be obtained anywhere in the area covered by the laser. Earthmoving equipment fitted with appropriate sensors can also be operated to control the level of excavation or filling with minimum input other than by the machine operator.

12.5 Setting out floor levels

A carpenter's spirit level should not be used for setting out the level of anything more than incidental work. It is not sufficient, in conjunction with a straight edge for instance, for getting a floor screed uniformly level. It is difficult to get concrete floors uniformly level to an accuracy better than 5 mm, and a contractor should always be warned when greater accuracy than this must be obtained with concrete. Usually discrepancies of 5 mm can be taken up in the floor screed of granolithic or terrazzo ground down to the desired smooth finish. To get tiling accurately laid, small pieces of tile are mortared onto the floor base at intervals across it, their level being fixed precisely to the correct finished level by use of the instrument level. A straight edge is then used to keep the finished tiling at the right level between tile pieces, which are cut off as the work proceeds. There are, however, some experienced tradesmen who exhibit astonishing skill in tiling an area perfectly level given only one level point.

12.6 Site investigations

Site investigations taken at an early, feasibility stage of a project will seldom be adequate for construction. More site tests will be necessary for individual foundations, etc. British Standard BS 5930: 1999, *Code of practice for site investigations*, acts as a general guide for further site tests, but this needs to be supplemented by information contained in other publications as suggested at the end of this chapter. The resident engineer will be expected to have an understanding of the major principles and techniques of soil mechanics so that he can direct work intelligently. But for specifying tests and interpreting their results, an experienced geotechnical engineer is essential, otherwise misleading assumptions can be made which later lead to serious trouble on a job.

There is an 'art' as well as a science in deciding what additional site investigations should take place when construction is started. Advice from a geotechnical engineer or engineering geologist should always be sought, but when choosing where to site extra boreholes or trial pits 'hunch' and 'suspicion' can play a part. A hunch should not be dismissed as unscientific; it can arise from studying the known facts and an apprehension that more needs to be known about some aspect of a situation than is currently known at the time. An experienced engineer will always worry more about what he does not know about below-ground conditions, than what he does know. Thus investigating some suspicion there might be a possible unconformity in conditions below ground can sometimes prove more revealing than gridding an area with boreholes at regular intervals – but not always!

12.7 Trial pits

Hand-dug trial pits are expensive, take time to excavate and are not always as informative as expected. They do, however, expose a formation so that it can be examined in detail. This may be important if thin layers of weak clay or pre-existing shear zones are suspected below ground. The starting size for a pit depends on the depth it is to be sunk. If required to a depth of 5 m for instance, it will have to be started between 3 and 3.5 m square, because the supports to it will have to be 'brought in' twice, and the reduced area at the bottom of the pit must be large enough for the men to work in, with a crane skip present and also possibly a pump.

Before starting a trial pit it is necessary to decide the depth and information to be sought and whether other means, such as augering or a borehole, would produce the information quicker and at less cost. If the requirement is simply to find rock level, or to ascertain whether soft material lies below hard (such as a boulder), a boring may be a better option. If one is looking for clay, silt or soft material, a most important matter is whether the pit is to find the full depth of such material or just penetrate into it. The former can be much more difficult and expensive than the latter and may prove impossible without groundwater lowering.

If undisturbed samples are to be taken it is necessary to know whether they are to be taken horizontally into the sides of the pit, or vertically from the bottom. Pushing a 100 mm diameter sampling tube horizontally into the side wall of a trial pit often involves the use of jacks, and digging the tube out is no easy matter.

12.8 Exploratory holes

Exploratory holes can generally be classified into three kinds:

- rotary core drilling by diamond drill to obtain samples of soil and rock;
- cable percussion driven lined holes in soft ground, sunk by clay cutter or shell;
- uncored holes drilled by rotary percussion drill in hard ground.

It is, of course, necessary to have an idea what sort of ground must be penetrated before the right type of investigation can be chosen; also it is necessary to know the kind of information required. It is not always possible to know the nature of the ground beforehand; soft ground can contain large boulders, and hard ground bands of soft or loose material. Mixtures of this type will cause delay.

Rotary core drilling

A rotary core drill uses a circular, diamond-embedded drill bit which cuts out a core of rock. The standard sizes in use are given in BS 5930. The most usual starting size adopted is 'H' (nominal hole diameter 99 mm) to give good sized cores of 76 mm diameter which are less liable to fracture during the cutting process and which permit size reduction to deepen a borehole. It is important that cores are inspected immediately upon withdrawal in order to note whether fractures are fresh and caused by drilling, or whether they are natural to the rock. The recovery percentage must be checked and recorded and may indicate the need for a change in equipment or technique. The cores must be labelled 'top' and 'bottom', the depth must be marked on them, and they must be placed for safekeeping and later inspection, in sequence, in purpose-made core boxes. A label should be attached to the box stating the borehole reference, date of start of drilling, etc.

When drilling, the need to get complete and reliable information on the groundwater is important. The water level at the beginning and end of each day's work should be measured, and preferably before and after each break or stoppage for testing. The sinking of the hole disturbs the natural groundwater conditions, but the changes in level recorded give valuable information on the probable natural conditions and the rate of inflows and outflows at various levels. On completion of a hole it is valuable to install a piezometer by which the longer term natural fluctuation of the groundwater levels can be recorded.

Particular attention should always be paid to any hole which the driller reports as difficult to sink – the drill bit gets jammed or the drill goes off line, or the hole has to be abandoned. Any of these can be the sign of a geological fault, unconformity of strata, a change of inclination of strata and so on. It is surprising how often one finds the drilling records for the cutoff of an old dam show a borehole missing in the very area where trouble is later experienced. So if a boring has to be abandoned it can be important to sink another one very close by, perhaps using a different technique for core recovery.

Light cable percussion drilling

Light cable percussion driven lined holes in soft ground are usually of larger diameter than rotary drilled holes, often 150 mm diameter to allow U100 samples to be taken. A deeper hole may need starting off at a larger diameter. The hole is excavated by bumping a 'shell' or clay cutter on the base of the hole. The shell is used on non-cohesive soils (e.g. sands and gravels), and is a heavy cylindrical tube with a lower cutting edge and some form of non-return flap valve inside. Material entering the shell is retained and withdrawn with the shell, which is removed every 0.5 m or so of boring and emptied for examination. The clay cutter is similar to a shell, but has a retaining ring at the base to hold the clay in, and has open slots either side for removal of the clay. The material inside the shell or clay cutter is partly disturbed but its nature can be inspected and logged. To take an undisturbed sample a 100 mm diameter sampling tube attached to rods is pushed or driven into the base of the boring, given a slight twist to break off the sample and withdrawn. Alternatively a down-the-hole hammer can be used to drive the tube. The sampling tube has a detachable cutting shoe with a small internal lip to retain the sample.

If the ground is very weak it may be necessary to push temporary lining down as the hole is deepened. After this it may be necessary to use a shell or cutter of slightly smaller diameter to continue drilling.

Percussion drilling

A percussion drill may be used to penetrate rock or boulders if no cores are required. A percussion chisel, usually of cruciform shape with a string of tools to give it weight, all suspended on a wire rope, is raised and let fall repeatedly on the rock base of the hole. The chisel has to be let fall with a clean blow on the base, and it is caused to rotate a little with each blow by the suspension wire having a left hand lay, and a friction grip attachment which lets the wire re-set from time to time. The drill chisel must be sharpened regularly. The rock chips are removed by water flush in small holes; in larger holes a bailer, very similar to a shell, has to be lowered at intervals to collect the chippings.

Sometimes it is only necessary to find the depth at which hard material, such as rock exists, or to drill sufficiently far into rock to ensure it is not a boulder overlying soft material below. In such situations a down-the-hole hammer drill can be used at a relatively small diameter. The blows to the cruciform bit are applied by a compressed-air operated hammer adjacent to the bit. The rock fragments are either blown out to the surface, or washed out by drilling water supplied through waterways inside the drill rods and can be examined. Small percussion drills of this type, 50–75 mm size, are fast and can penetrate something of the order of 6 m of rock or concrete per hour.

12.9 Other means of ground investigation

The hydraulic digger (or backhoe) is useful as a means of revealing the nature and extent of shallow overburden material on site. It can excavate a trench up to 3–4 m deep in soft or moderately hard material, at a fast rate and cheaply. The substantial cross-section of material then revealed for inspection can be more informative than samples from a few borings. However, trenches of this depth must be securely supported before access for inspection is allowable. A sampling tube can sometimes be pushed into the base of the trench using the digger bucket, and then dug out by the same machine.

The auger may be used for boring holes in soft materials. A lining may be required to keep a hole open during and after boring. Large augers, machine driven, are used for sinking shafts for the formation of in situ concreted piles. For site investigations, the hand-auger is a simple little tool, usually of 75–150 mm diameter, for penetrating shallow depths of soft material. About 300 mm of material is penetrated at a time before the tool is withdrawn and the material taken out of it and examined. Two men are usually required to twist the auger, the hole being watered from time to time if necessary in order to reduce friction. Penetration is usually of the order of 1.5–2.5 m; to get a hole deeper than 3 m the ground has to be very soft. Gravel or cobbles cannot be penetrated. The tool is useful for locating the extent and depth of shallow, very soft overburden material.

As an alternative to rotary core drilling the same rig can be used for open hole drilling where the drill bit cuts all material within the hole. Casing may be needed in unstable ground. This can be a rapid means of reaching a required depth to carry out a test or install instruments.

12.10 Judging the safe bearing value of a foundation

The safe bearing value for a given foundation material ought not to be decided by the resident engineer but by the engineer or his specialist advisers. However, the engineer will not thank the resident engineer for referring to him questions about foundation materials which are obviously satisfactory, such as gravel and rock, where the load thereon is well within the traditionally accepted bearing strength. Standard field descriptions are given in BS 5930 for various materials and BS 8004, Table 1 shows allowable bearing values for such materials for preliminary design purposes.

In clays or silts, or materials having clay bands or organic layers, and other mixtures containing weak layers, special investigations, sampling techniques and sophisticated analyses may be necessary before a safe bearing value can be advised – dependent upon the type of structure the formation is to support. These matters need to be considered by an experienced geotechnical engineer. Site tests, such as the 'standard penetration test', vane shear tests and permeability tests, may be used but these must be regarded as an adjunct only to more sophisticated investigation techniques. Details of the standard penetration test are given in BS 1377 Part 9: 1990, para 3.3. Its widest use on site is to reveal any weak spots in an otherwise consistent foundation material.

12.11 Testing apparatus for a site soils laboratory

The usual apparatus suitable for a small soils laboratory on site, to be run by the resident engineer's staff after proper instruction from a geotechnical engineer, is set out below.

For moisture content determinations

1. Beam balance weighing by 0.01 g divisions.
2. Drying oven, thermostatically controlled. (Not absolutely essential. For rough measurement of moisture content the sample can be dried on a flat tray over a stove.)
3. Six drying trays.

For grading analyses of soils

4. A set of BS sieves (woven wire) with lid and pan for each different diameter:
   (a) 300 mm dia – 38, 25, 19, 13, 10 mm. (These can also be used for testing concrete aggregate gradings.)
   (b) 200 mm dia – 7, 5 and 3 mm, and Nos. 7, 14, 25, 52, 72, 100 and 200.
5. Balance weighing up to 25 kg.
6. Balance capable of weighing up to 7 kg by 1 g divisions.

For in situ density test (sand replacement method)

BS 1377 Part 9:1990 gives four tests of which Test 2.2 is the most useful because it can be used on fine, medium and coarse grained soils. A metal tray with a 200 mm diameter hole cut in it is placed on the formation and material is excavated via the hole. The volume of the excavation is measured by pouring uniformly graded sand into it whose bulk density has been measured.

Apparatus required (additional to 1, 2, 3 and 6 above):

7. Small tools for excavating hole.
8. A rigid metal tray 500 mm square or larger with a 200 mm diameter hole cut in it.
9. Dried clean sand all passing No. 25 sieve but retained on No. 52 or 100 sieve and suitable airtight containers for storing it. (About 20 kg of this sand will be required initially.)
10. A pouring cylinder (as BS 1377 Part 9 Fig. 4).
11. A calibrating container 200 mm diameter by 250 mm (as BS 1377 Part 9 Fig. 5).
12. Air-tight containers for the excavated soil.

The method can be applied to larger test holes in soils containing some gravel; the sand being poured in layers from a can with a top spout. A length of hose is attached to the spout with a conical tin shield wired to the lower end, so the sand has only a short standard free fall. Tests to fill measured containers can show the accuracy in ascertaining the bulk density of the sand as poured.

For compaction tests

BS 1377 Part 4: 1990 describes tests using 2.5 or 4.5 kg hammers on soils with or without coarse grains.

For the 2.5 kg test on fine and medium grained soils the apparatus required (additional to items 1, 2, 3 and 6 above) is:

13. Compaction mould (BS 1377 Part 4 Fig. 3).
14. 2.5 kg metal rammer and guide (BS 1377 Part 4 Fig. 4).
15. Palette knife.

16. Glass sheet or metal tray (for mixing in added moisture to sample).
17. 19 mm sieve (from Item 4 above).
18. A 1-litre glass graduated measuring cylinder (for measuring volume of surface-wet material over 19 mm sieved out).

The compaction test and in situ density test are important for earthworks construction. The former indicates the maximum density and optimum moisture content of fill material achieved under 'standard' compaction; the latter shows the density of fill achieved. Specifications often require fill as placed to achieve 90 or 95 per cent of the maximum density obtained under one of the compaction tests stipulated; the 4.5 kg hammer test being used for road construction, the 2.5 kg test for other earthworks. For road works a CBR test is often essential to check design assumptions for the strength of sub-grade. For accuracy this is normally carried out in the laboratory using standard equipment as set out in BS 1377 but in situ tests can also be done as a ready check on soil strength.

A small unconfined compression testing apparatus for testing the shear strength of 38 mm diameter undisturbed clay samples is a useful addition to the site laboratory in certain circumstances. This machine is cheap, easy-to operate and gives a useful indication of variations of clay strength, as for road making, etc. The results given by it are not, however, adequate for design purposes. The triaxial compression testing machine would be used for testing soils for design purposes; but this is a sophisticated piece of apparatus, not suitable for site control purposes unless a full-scale soils laboratory has been set up on site under the direction of a properly qualified geotechnical engineer. It is also useful to have some standard 100 mm diameter sampling tubes on site, to prevent a delay in getting such tubes when an excavation reveals material that needs to be tested.

Provided proper briefing has been given by an experienced geotechnical engineer concerning the techniques of testing to be followed, the foregoing apparatus should permit a useful range of quality control tests to be carried out on site. Most other tests that might be required, such as consolidation, permeability and triaxial compression tests, must be regarded as advanced laboratory tests to be carried out by trained technical staff.

12.12 Site layout considerations

Haulage roads

The roads within the site have to be planned and designed by the contractor. Lengthy roads may be required to take excavated material to a dumping ground, or from a borrowpit to the construction site. They exclude any traverse over public roads. The design of such roads is related to the type of excavating machinery the contractor proposes to use. Motorized scrapers and balloon tyred wheeled loaders can pass over hard to moderately soft ground and will not seriously disrupt the surface. In most cases roads have to be designed for haulage trucks, which can impose heavy wheel loads when laden. Roads for them must be of adequate thickness, suitably topped (or constantly regraded) to prevent rutting and ponding, and well drained. Any attempt to save money by building an access road of inadequate thickness, without proper drainage ditches either side and a surface kept to a camber to shed rainwater, is a false economy. It will quickly break up and cause repeated delay to the job. Flat tracked machines can pass occasionally over metalled, waterbound, or sprayed and chipped roads without causing much damage. Machines with gripped tracks, such as large dozers, will quickly break up the surface of any road.

The consequence of the foregoing is that internal roads on site have to be designed according to the anticipated usage of them. For maximum output from motorized scrapers it is important that haulage roads should have easy gradients. Laden haulage trucks are also frequently slow on steep gradients, both uphill and downhill. Mud is a particular nuisance when trucks have to go on public roads; frequent cleaning of the road and hosing of traffic leaving the site will be needed if public objection is to be avoided.

Planning bulk excavation

The order in which an excavation is to be undertaken has to be planned. The excavating machine must be able to work to its maximum capacity attended by a continuous flow of dump trucks in and out. As bulk excavation proceeds, formation trimming and minor excavation will follow, then the placing of fill or concrete. For speed of execution these follow-on operations will need to be started before the bulk excavation is completed. Hence the excavation must be planned in such a manner that the different operations carried on simultaneously do not interfere with each other, and that excavating machines can withdraw without difficulty after their work is completed.

Concrete production plant

This needs to be positioned to give easy delivery to the parts of the work where the main concrete is required. Delivery lorries to the stockpiles of aggregates should preferably not follow the same routes as muck-shifting plant, or they will pick up mud and track it into the aggregate bays. The bays should have concrete floors laid to a fall so the aggregate can drain.

Power generators and compressors

These may need to be housed, even if mobile, because their noise can create a nuisance to local residents. Their siting should be such that the noise they create is 'blanked off' from any residences. Even though the noise from a construction site may be 'music to the ears' of a civil engineer who likes to hear the job 'humming along', the public at large take a diametrically opposite view. Authorization of night working may be difficult to obtain if attention is not paid to reducing noise as much as possible.

Extra land

Extra land outside the site or extra access to the site can be obtained by the contractor if he so desires, provided it is not disallowed under the contract, and the contractor gets the necessary permissions, wayleaves, etc. and bears all costs involved.

Main offices

The contractor's main offices and stores need to be near the site entrance. Most vehicles carrying materials to site must stop at the checker's office, and it is convenient to have this near the agent's offices and the stores. The resident engineer's office should not be far away from the agent's offices, so that easy communication is maintained at all staff levels, and there is economy in providing telephone, heating, lighting and sewerage.

12.13 Temporary works

Temporary works are mostly designed by the contractor, but the resident engineer will need to review the design because of the safety responsibility also held by the engineer (see Sections 10.2 and 10.3). On a large project the temporary works may comprise major structures such as caissons, cofferdams, river diversion works, sheet piling, access bridges, etc. Designs for such structures will normally have to be forwarded to the engineer for his consent, though the resident engineer should apply his site checks first so that he can draw the engineer's attention to any matters of doubt seen on site.

12.14 Work in public roads

Under the New Roads and Street Works Act 1991 (applying to England, Wales and Scotland) a **street works licence** has to be obtained from the relevant highway authority before any work to install, maintain or alter apparatus in the highway is permitted, except emergency work. The ICE Conditions of

Contract state that the employer will obtain this licence, but that the contractor must comply with all other requirements of the Act and any conditions attached to the licence (Clause 27). The contractor has to give notices as required by the Act and the Street Works (Registers, Notices, Directions and Designations) Order 1992. Notice to start work must be given at least 7 working days in advance, and the work must be started within 7 days of the notified starting date. The Street Works (Qualifications of Supervisors and Operative) Regulations 1992 require a qualified supervisor and at least one qualified operative to be full time on site. Road reinstatement requirements are set out in the Street Works (Reinstatement) Regulations 1992 and a Code of Practice issued by the Secretary of State.

A highway authority can direct the timing of works, require safety measures and stipulate avoidance of unnecessary delay or obstruction. A standard charge for inspections can be made and the authority can also charge for the occupation of a highway where works are unreasonably prolonged, and for the cost of temporary traffic regulation.

The highway authority is required to keep a register of street works; this can be of use to the contractor but, in the nature of things, it may not show every service that lies underground nor provide exact information as to its position. A highway (termed a 'street' in the 1991 Act) normally means all the land between the boundaries of private properties fronting on a public road, that is, including the road verges.

The diversion of existing services often requires joint action by the agent and resident engineer. If need be, the resident engineer should arrange for meetings with the district engineers of the authorities concerned, for example, county or district roads department, gas, water, sewerage, electricity, telephone and TV cable authorities. The resident engineer must see that the reasonable requirements of the various authorities are complied with by the contractor; on the other hand he should help to resist any unreasonable requests being put upon the contractor. Most authorities prefer to divert their own services; many will not permit a contractor to undertake diversion of their equipment. Similarly with respect to final road reinstatement, the road authority has power to do this and may prefer to do so. A common requirement of a road authority is that a trench for a pipeline, sewer, etc. laid along a road must be at least 1 m away from the road edge (i.e. fully in the road or fully in the verge), except where it has to cross below a road edge.

12.15 Site drainage

Difficulty often occurs in draining a site where large scale earthmoving is taking place. The excavations disturb the natural drainage of the land and large quantities of mud may be discharged to local watercourses during wet weather. Complaints then arise from riparian owners and water abstractors downstream. If this possibility should occur the resident engineer should advise the contractor to approach the appropriate drainage authority (the Environment Agency in England and Wales) to seek advice on the best course of action to alleviate the problem, such as arranging some form of stank to pond the runoff and allow the heaviest suspended solids to settle out. It is the contractor's responsibility to dewater the site, and this includes the obligation to do so without causing harm or damage to others.

Dewatering can range from simple diversion or piping to ditches, to full-scale 24 h pumping and groundwater table lowering. It is usual to cut perimeter drains on high ground around all extensive excavations. In dry weather this may seem a waste of time, but once wet weather ensues and the ground becomes saturated, further rain may bring a storm runoff of surprising magnitude. If no protection exists for these occasions extensive damage can be caused to both temporary and permanent works. The resident engineer should assist the contractor to appreciate the danger of flood damage by providing him with data showing possible flood magnitudes. A frequently used precaution is to assume that a flood of magnitude 1 year in 10 (i.e. 10 per cent probability) will occur during the course of construction.

The need to dewater an excavation in the British Isles is the rule rather than the exception. Once dewatered an excavation should be kept dewatered. To repeatedly dewater an excavation during the day and let it fill up overnight can cause ground instability, and timbering to excavations may be rendered unsafe. The need for 24 h pumping should be insisted upon by the resident engineer if he thinks damage or danger could occur from intermittent dewatering. The electric self-priming centrifugal pump is the most reliable for continuous dewatering, having the advantage that it is relatively silent for night operation as compared with petrol or diesel engine driven pumps.

For groundwater lowering, pointed and screened suction pipes are jetted into the ground at intervals around a proposed excavation and are connected to a common header suction pipe leading to a vacuum pump. It may take a week or more before the groundwater is lowered sufficiently, but when the process works well (as in silt or running sand) the effect is quite remarkable. It permits excavation to proceed with ease in ground that, prior to dewatering, may be semi-liquid. However, it can be difficult to get the well points jetted down into ground containing cobbles and boulders; and in clays the well points need to be protected by carefully graded filters, or the withdrawal of water may eventually diminish because the well point screens become sealed by clay.

Special precautions must be taken to avoid damage to any adjacent structures when dewatering any excavation or groundwater lowering. In some soils groundwater lowering may cause building foundations to settle, causing considerable damage. The contractor may have to provide an impermeable barrier between the pumped area and nearby structures, monitor water levels and perhaps provide for re-charge of groundwater under structures. A vital precaution is for the resident engineer to record in detail all signs of distress (cracks, tilts, etc.) in adjacent structures and take photographs of them, dated and sized, before work starts, in order to provide evidence of the extent of any damage which may occur.

The drainage of clay or clay and silt can present difficulty. The problem is not so much that it cannot be done, but that it can take a long time, perhaps many weeks. Sand drains (i.e. bored holes filled with fine sand), can be satisfactory as part of the permanent design of the works, but they usually operate too slowly to be of use during construction. If ground is too soft, any attempt to start excavating it by machine may make matters considerably worse, and end with the machine having to be hauled out. The act of removing overburden may make a soft area even softer as springs and streams, otherwise restrained by the overburden material, break out and change the area to a semi-liquid state. If the resident engineer sees the contractor moving towards these difficulties he should advise him of the possible consequences, and endeavour to give assistance in devising a better approach. A paramount need may be to call in an experienced geotechnical engineer to investigate the problem and give advice as to the best policy to handle the situation.

References

Schofield W. *Engineering surveying.* Butterworth-Heinemann, 2001.

ICE *Site Investigation in Construction*, 1993. (Part 1 Without site investigation ground is a hazard; Part 2 Planning, procurement and quality management; Part 3 Specification for ground investigations; Part 4 Guidelines for the safe investigation by drilling of landfills and contaminated land.)

Clayton C.R.I., Simons N.E., Matthews M.C. *Site investigation.* Blackwell, 1995.

Transport and Road Research Laboratory, Road Note No. 17, *Protection of Subgrades and Granular Bases by Surface Dressing.* (An early publication but useful for construction of site roads.)

BS 1377:1990: *Methods of test for soil for civil engineering purposes*, Parts 1–9. (Part 1 General requirements and sample preparation; Part 4 Compaction tests; Part 9 In-situ tests.)

BS 5606:1990: *Guide to accuracy in building.*

BS 5930:1981: *Site investigations code of practice.*

BS 5964:1980: *Methods for setting out and measurement of buildings.*

BS 8004:1986: *Foundations code of practice.*

13

The resident engineer's office records

13.1 Records and their importance

The records a resident engineer (RE) must keep are essential for deciding what payments are due to the contractor and what claims and extensions of time are to be allowed. They record the quality of the works as built and all tests thereon. They keep track of progress and decisions made, the financial expenditure and the probable final cost of the job. When the project is completed they provide details and drawings of all the works as built.

The records required come under four categories:

- *historical* – weekly reports showing progress of the work, a diary, and minutes of all discussions and decisions relating to progress and changes to the work, and daily weather data;
- *quantitative and financial* – measuring quantities of work done, recording facts about varied work and contractor's claims, estimating expenditure to date and probable final cost of works;
- *qualitative* – recording all observations of the quality of the work and materials used, including nature of foundation materials, test results, test certificates from manufacturers and suppliers; performance tests on completed works;
- *'as built' records* – record drawings of all the works as built, details of the origin and quality of all materials used in the works, names of suppliers, manufacturers' instruction manuals for equipment, and the operational instruction manual for the works as a whole.

13.2 The correspondence filing system

A correspondence filing system of the type outlined below is needed.

General files (Series 1–9)

1. Employer (including copies of letters sent by the engineer to the employer).
2. Monthly progress reports to employer – drafts sent to the engineer; and copies as sent by the engineer to the employer.
3. Meetings file – second copy of notes of meetings on site with employer, engineer, or others, in date order (from files 1, 10 and 11).
4. Planning authorities, etc.
5. Road authorities and public utilities.
6. Miscellaneous, for example, re telephone, visitors to site, etc.
7. Staffing – re appointment of inspectors, office staff, etc.
8. Miscellaneous (personal) – RE's personal correspondence relating to the job.

Head office (Series 10–19)

10. Engineer – correspondence with.
11. Specialist advisers – correspondence with other advisers (e.g. geologist, landscape architect, etc.).
12. Informal memos to designers – copies of notes sent to colleagues in the engineer's office (though most correspondence should be through the engineer).
13. Test certificates, etc.

14, 15, …, etc. Other special subjects as required.

Separate supply contracts and sub-contractors (Series 20–29)

20. Supply contractor A.
21. Supply contractor B.
22. Supply contractor C, etc.
23. Nominated sub-contractors/suppliers.

Main contractor (Series 30–39)

30. Contractor's head office – copies of letters sent by the engineer to the contractor (the RE would not normally send any).
31. Contractor's agent – excludes Instructions File 32.
32. Instructions to contractor.
33. Applications for interim payment from main contractor (a bulky file).
34. Engineer's interim certificates and correspondence thereon.

35/1. Variation Orders – Passed (copies as signed by engineer).
35/2. Variation Orders – Drafts as sent to engineer, or Pending.
36. CVIs from contractor (see Section 13.3).
37/1. Claims – Pending or Rejected.
37/2. Claims – Settled.

Under file 37/1 all claims should be numbered, usually in order of receipt. Each claim and correspondence thereon (copied from file 31) should be filed as a separate section with a tabbed index. All claims should be listed at the front of 37/1 and marked with the Subject; and whether 'Pending', 'Rejected', or 'Settled' (in the last case transferred to File 37/2).

The above list is only an example for guidance. If need be any file can be broken down into sub-files, for example, 11/1, 11/2, etc. On a small job some files can be merged; on a large job a more extensive system may be required. It is best to file correspondence under the name of the addressee. This way letters do not get mis-filed. Important letters that take time to deal with should be photo-copied so that the original can go straight on file and be kept there, and the copy can be used for working purposes.

Several files will be required for dayworks sheets. They pose a special problem which is dealt with in Section 13.9.

13.3 CVIs from contractor and instructions to contractor

CVIs are **confirmation of verbal instructions** (sometimes called 'CVOs' – confirmation of verbal orders) sent by the contractor to the RE when some verbal instruction has been given to the contractor as described in Section 17.6. These CVIs raise a number of problems. They should be filed together with the RE's reply and subsequent correspondence. Usually the contractor will have numbered the CVIs in order. If a dispute arises as to whether any payment is due under a CVI, this should be given a claim number and transferred to the Claims File 37/1.

As already mentioned all instructions to the contractor have to be given in writing or, if given orally, confirmed in writing as soon as possible. Major instructions will be by letter, copies being put on File 31; but there are many other day-to-day instructions of a minor nature, such as supplying the general foreman with a sketch of some levels, etc. To deal with these the RE should be supplied with forms that can be used for such instructions. The forms should be A4 size, simply headed thus:

A-B-C project
Site instruction No .... Date ....

To: The Agent, Messrs XYZ
From: Resident engineer

Subject ....

You are requested ... etc.
....
....
This instruction does/does not constitute a variation of the works.

Plate 1a. The hydraulic model for the Ghazi Barotha Hydropower project, Pakistan, 1994

Plate 1b. Power channel inlet structure, Ghazi Barotha Hydropower project, Pakistan 1999

Plate 2a. An early use of a large tunnel mole for driving the 11.0 m diameter hydropower supply tunnels on the Mangla dam project, Pakistan, 1963

Plate 2b. A Euclid R60 522 kW dump truck, 36 m^3 heaped capacity. VME Construction Equipment GB Ltd, Duxford, Cambridge, UK

Plate 3a. A Terex TS24C 552 kW twin-engined 36 m^3 capacity scraper for earth-moving. Terex Equipment Ltd, Motherwell, Scotland

Plate 3b. A Kato 162 kW tracked excavator, with 1.6 m^3 bucket. ACP Holdings, Leicester, UK

Plate 4a. Rollers compacting side slopes of power channel; filter layers and concrete to follow. Ghazi Barotha Hydropower project, Pakistan 1999

Plate 4b. A badly rutted formation, probably due to excessive moisture content of the fill

Plate 5a. Ten metre diameter steel can to form hydropower penstock, Ghazi Barotha Hydropower project, Pakistan 2001

Plate 5b. It can be easier to get out of this sort of trouble than to decide whether the conditions 'could not reasonably have been foreseen by an experienced contractor' under Clause 12 of the ICE conditions

Plate 6a. A problem for the designer under the Construction (Design and Management) Regulations 1994 who has to pay 'adequate regard to the need to avoid foreseeable risks' (see p. 107 *et seq.*). The designer may decide a stable berm should be cut into the hillside before the pipe trench is excavated

Plate 6b. Steel piling sometimes does not go where the civil engineer thinks it has – especially when he was not aware the ground contained boulders or other obstructions

Plate 7a. Erosion of a newly placed embankment by rainfall runoff from the formation above. This is likely to be a contractor's risk, as it is 'a condition due to weather conditions' under ICE conditions Clause 12

Plate 7b. An early photo showing unsafe conditions not now permissible. The designer would now have to consider any safer alternative route for the trench, or require a safe berm to be cut into the hillside first to give a shallower trench. The contractor must fully support the trench sides and require his men to wear hard hats.

Plate 8a. Presumably somebody thought this trench was stable – and was proved wrong

Plate 8b. The ubiquitous backhoe loader used on many sites. That shown is the JCB 3CX 56.5 kW, with shovel up to 2.3 m wide and hoe bucket 0.3–0.9 m wide. JCB Ltd, Rocester, UK

Plate 9a. A Mastenbroek 17/17 trenching machine with variable-offset heavy-duty cutting chain for trenches up to 0.6 m wide by 1.8 m depth. Larger machines are made. J. Mastenbroek & Co. Ltd, Boston, UK

Plate 9b. A properly supported trench

Plate 10a. It is almost universal to use the hydraulic excavator also as a crane

Plate 10b. Well-designed steel shuttering for a single-lift wall pour of concrete. Safety regulations require full boarding and a toeboard to the access walkway

Plate 11a. An early photo of congested pipework being built in the base of a shaft, taking more than a week to complete. Hard hats must now be worn.

Plate 11b. A sophisticated gantry for handling and placing pre-cast concrete bridge and other units. Ghazi Barotha Hydropower project, Pakistan 2001

Plate 12a. On any job not properly supervised the strangest things can happen. It looks as if someone forgot the reinforcement until after the first placement of concrete. A resident engineer would stop the work and require it to be redone properly

Plate 12b. Judging the quality of a concrete mix – a satisfactory mix. Photos from *Concrete Materials and Practice* by LJ Murdock, KM Brook, JD Dewar, 6th edition, Edward Arnold, 1991

Plate 13a. Judging the quality of a concrete mix – an over-sanded mix. Photos from *Concrete Materials and Practice* by LJ Murdock, KM Brook, JD Dewar, 6th edition, Edward Arnold, 1991

Plate 13b. Judging the quality of a concrete mix – a mix that is too harsh. Photos from *Concrete Materials and Practice* by LJ Murdock, KM Brook, JD Dewar, 6th edition, Edward Arnold, 1991

Plate 14a. The typical fluidity of pumped concrete; but it still needs vibrating in place

Plate 14b. Reinforcement must be accurately dimensioned and bent, and laps must be provided that permit the steelfixer to make adjustments for unavoidable discrepancies in the concrete work

Plate 15a. Column base excavations are always liable to fill with rainwater. A sump should have been excavated beside each to permit dewatering before concreting

Plate 15b. Sand runs on a concrete wall surface

Plate 16a. If reinforcement is not designed to provide for easy access this is what happens to it. Nor should the platform be used for access with uneven gaps, no railing or toe boards, etc.

Plate 16b. Accurately placed reinforcement, strong enough to walk on

Plate 16c. Concrete which needs to be cut out: honeycombing and apparently inadequate cover to reinforcement

If an instruction does not vary the works or does not entitle the contractor to varied or additional payment, this should be stated. If the instruction does vary the works, a statement should be added indicating how it is to valued: for example, at dayworks rates; or bill rates; or at varied rates to be proposed by the engineer. If payment rates will have to be discussed with the contractor, the note can simply say 'at rates to be agreed'. A copy of each instruction must be kept on file. They can be handwritten.

There is no need to issue an instruction to the contractor if, for instance, an inspector or assistant engineer finds some formwork out of line and orally requests it to be corrected. Matters requiring compliance with the workmanship specified are only oral restatements of the contract requirements. If any such oral request results in a CVI from the contractor which implies a claim will be submitted for extra payment, the CVI should be promptly countermanded by a formal letter, rejecting any basis of claim.

Dimensions given orally should, for clarity, be confirmed by a written memo.

13.4 Register of drawings

Drawing registers are required:

- one should be for 'Drawings Received' logging – title; reference; who from; date received; size and type; how many copies received;
- the other should be for 'Site Drawings', that is, drawings made on site, logging – consecutive site reference number; subject/title; size and type; to whom copies sent and when.

Logging size and type of drawing is useful when trying to find a drawing.

13.5 Daily and other progress records

The principal records that have to be kept in this category are:

- the inspectors' daily returns;
- the site diary and weather records;
- the RE's diary;
- weekly and monthly progress records;
- progress charts (these are dealt with in Chapter 14).

The inspectors' daily returns are a vital record. If no inspectors are employed, then each assistant engineer should complete a data form for his own section. If the RE is on his own he should endeavour to keep the necessary log going himself. A typical inspector's return is shown in Fig. 13.1. The sheet is purposely simple because it hopes to encourage the inspector to

Date 3/8/01
Shift: Night/Day

INSPECTOR'S REPORT

	WORK DONE BY CONTRACTOR (1) Valve Tower and Bellmouth, (2) Tunnel and Stilling Basin, (3) Dam, (4) Quarry and Road, (5) Road Diversions, (6) Miscellaneous	Resources
3	1 D8 dozing Rip-rap limestone up U/S slope east side. 2nd D8 and roller spreading sand in northwest corner. D6 and roller also spreading sand, south east of wall. Cat 953 dozing sand up south west hillside to make up gradient to original slope, before placing filter. Small JCB 406 assisting D8 in north west corner. Rollers compacting fill both sides of wall. Leyland truck and bowser watering bank. 2 Aveling RDs and 1 Cat D25 trucks and 2 Cat 621 scrapers delivering sand until 2.30 pm.	1 Foreman 1 Ganger 1 Chargehand 2 D8 + drivers 1 D6 " 1 Cat 953 " 1 JCB " 2 Roller " 1 Leyland " 1 Bowser " 2 Aveling " 1 D25 2 Cat 621 " 10 Labourers
4	2 O e K RH30s and JCB 425 loading vehicles until 2.30 pm. Drillers drilling quarry face, Shotfirer and 1 labourer blasting.	1 Ganger 1 Chargehand 2 O + K 1 JCB 2 Drillers 1 Labourer
5	Poclain tidying up bermes at side of Green road. Hired lorry removing excess spoil to tip, D4 levelling bermes at side of road. 13 labourers working at Middle road tidying limestone bermes and 5 finished at 12.30 pm. 13 labourers working at Green road tidying up verges, helping curblayer to make curbs and channels up to Gully gratings.	3 Gangers 1 Poclain + driver 1 D4 " 1 Lorry 1 Joiner 1 Joiner 1 Curblayer 26 Labourers
6	Men felling and burning trees in Hawthorne wood 10 finished at 12.30 pm.	1 Ganger 16 Labourers
	REPORT ALL DELAYS AND BAD WORK 1 D25 broken down all-day 1 D8 " " " 1 Daf truck broken down from 9 am. 1 D6 tractor " all day vibrating roller and JCB 2X standing all day	Time lost and No. of men Involved 1 D8 driver 1 truck "

Hours worked by Inspector 9½

Fig. 13.1. A well-written inspector's report

put down the following information:

- where and what type of work was being done that day;
- how many men and what machines were engaged;
- what delays and bad work were experienced and why.

The inspector is given an aide memoire at the top of the form to remind him of the separate sections of the job needing coverage. The last section permits the inspector to comment on what he sees, in addition to logging down delays, etc. This gives the inspector an opportunity to put his comments on record, thus contributing to the RE's successful control of the job.

Inspectors' records are invaluable for dealing with claims from a contractor for delay, disruption, lack of instructions, or 'uneconomic working'. These are all difficult to handle if only general progress charts are available.

The site diary is a day-to-a-page diary which notes matters not on inspectors' daily returns, such as weather, visitors to site, meetings held. Weather records are important, but need not be strictly meteorological. It can suffice to note weather which affects work, for example, stoppages due to rain or snow; freezing conditions that can affect concrete; showers interrupting concreting; excessive heat causing over-drying of concrete, etc. A note such as 'Heavy showers interrupting concreting' can, for instance, be the explanation for leaks found later in the concrete walls of a tank which appear to be due to poorly compacted concrete.

The RE's diary is not easy to keep. The aim must be to record events not recorded elsewhere, such as decisions on problems; comments made by the employer or specialist advisers; and important telephone calls. When things get in a tangle and misunderstandings occur, it can be particularly important to be able to say, with certainty, *when* a discussion or telephone conversation took place.

It depends on the style of operation of the agent how meetings with him are recorded. Formal meetings (usually over claims) have to be minuted and agreed. But many informal discussions will take place between an agent and the RE. It is not usual to minute these. A good agent will often discuss some problem with the RE; and if this leads to some oral agreement or instruction from the RE, the agent will not act otherwise. All the RE needs to do is to make a note in his diary of any matter decided.

Many a job is run almost entirely by oral discussion and agreement between RE and agent, without any need to record what was decided. However, when a complicated series of decisions has been agreed upon, a written list of these might be supplied to the agent so that the staffs of either side know what to do. In other unfortunate cases where an agent makes things difficult and takes every opportunity to lodge a claim, it may be necessary to confirm every instruction in writing.

The *weekly report* is commonly sent to the engineer. A typical example is shown in Fig. 13.2. A *monthly report* should be sent to the employer, primarily to inform him of progress to date. A draft of this is usually sent to the engineer, for him to amend as necessary and send under his own hand to the employer. On overseas sites, weekly reports are not usually adopted; instead a monthly report will go direct from the RE to the employer with copies sent to the engineer.

13.6 Quantity records

For admeasurement contracts, measurement of quantities of work done will be one of the most important tasks undertaken by the RE and his staff. Two essentials for any system are:

- it must be possible to ascertain from the records what has been measured and what has not been measured;

XYZ CONSTRUCTION SCHEME
Weekly Report No. 8
(For week ending 26th May 2001)

Contract 52: Extension to pumphouse

Continuous rain delayed waterproofing of roof until 25th. Plastering sub-contractors started work on 21st. They will finish next week. Installation of switchboard awaits arrival of manufacturer's fitters promised for 29th.

Labour: 2 plus 2 plasterers.

Contract 54: Electrical re-wiring

Cable laying completed but awaits connection to switchboard. Lighting wiring completed as far as possible. Contractor withdrew men from site on 24th and will return to complete after switchboard connection made.

Contract 57: Mainlaying

The extra sluice valve required at Ch. 3500 was delivered. This completes all deliveries under the valve contract (Contract 53).

Delivery of steel 600 mm diam. pipes continues. Total delivery to date 4600 m or 82 per cent order (Contract 55).

Supply of small cast-iron pipes and fittings continued. Only a few items now outstanding (Contract 56).

Approx. 260 m of pipe laid during week by two gangs working from point 'F' forward and point 'G' back; chainages 3260–3840 and 4480–4210 respectively. Suppliers are being pressed for delivery of special T-junction Ch. 3500 which is promised for next week.

The crossing under the two main-line tracks was completed successfully at Ch. 4400 and in 14 h continuous working.

Labour: 1 General foreman, 2 section foremen, 3 gangers, 3 jointers, 7 drivers, 11 labourers.

Plant: 2 JCB excavators. 1 Case loader, 3 dumpers, 2 D4 dozers, 4 lorries.

Testing. Testing to point F from H was satisfactory and the section left out at point F for testing was re-inserted and the line made good. Main has now been passed from Ch. 500 to Ch. 3260.

Miscellaneous. A director of Messrs. Smith accompanied by the sports ground manager inspected the reinstated trench through the sports ground and was satisfied.

General

Weather:	Wet. 2.15 in. rain	
	Temp. 57° to 68°F.	
Lost time:	2 men × 6 h	Contract 52
	Nil	Contract 57

Visitors

Director Messrs. Smith & Co.	25th
District road inspector to Contract 57.	25th
Representative pipe suppliers.	21st

A.B.C.
Resident Engineer

Fig. 13.2. Typical weekly report

- the records must clearly show what has been paid for (under interim certificates) as distinct from what has been measured.

The foregoing can be catered for by showing, in the quantities calculation book, two columns as shown in Fig. 13.3. The first column shows the total volume for part of the works; the second shows an estimate of how much has so far been certified for payment.

The measurement engineer will work from notebooks containing dimensioned sketches of the amount of work done measured by himself, or by assistant engineers and inspectors who send him notes. All these he will file, and armed with the quantities he calculates as a result, he checks the contractor's monthly claim in detail, item by item. Where he thinks something has been mis-measured or over-claimed by the contractor he raises this with the agent – or more usually with the contractor's quantity surveyor. Comparison of quantity measurements takes place to see where the difference might lie. If the difference is one of interpretation of how the quantity should be measured, the RE will

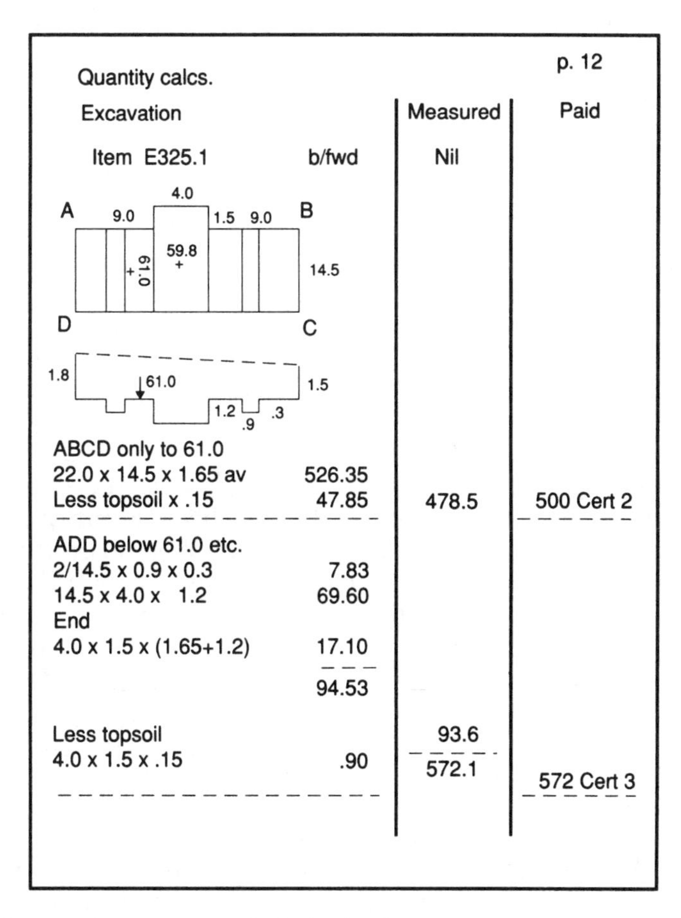

Fig. 13.3. A page of a quantities calculation book

have to decide what is the correct interpretation. If the contractor wishes to argue the matter it becomes 'a claim', and is put in the Claims Pending file with all details attached.

The RE may raise the dispute over interpretation with the engineer's contracts department when he sends in the contractor's application for checking. If the contracts department agree with the RE's ruling, there is no further action to be taken; but if the department agrees with the contractor they may re-insert the contractor's claimed amount or whatever the RE has advised would be the figure in that case.

Measurement sheets are best filed in 'lever-arch' files, in order of the Bill Items. In the early stages of the job it may not be worthwhile attempting to make any exact calculation of the quantity of an item. In that case a rough sketch is drawn on the measurement sheet, showing how an estimate of the work done has been calculated.

13.7 The contractor's interim payment applications

The contractor's monthly application for interim payment is usually set out as shown in Fig. 13.4. Initially only the items under which work has been measured need be listed, divided into bills. Later, when all the work listed on a page of the bill of quantities has been completed, the page total only need be quoted in the bill summary. Where work is incomplete and only an estimate can be made of the work done, the RE may accept the contractor's figures if they are reasonable. But where work is completed, the contractor's final measurement of a quantity should check reasonably with the RE's estimate. In such matters as excavation the RE should not require an unreasonable degree of accuracy, measurement to the nearest 1 m^3 is adequate. For minor differences between the contractor's figures and the RE's, a compromise figure may be agreed. Major differences should be looked into.

To aid estimating the probable final cost of the works (see Section 14.9) it is best if extra items are put on a separate sheet or sheets, grouped for each separate bill of quantities, where possible. Extra items not authorized by a previously issued variation order (VO) must be checked by the RE. If he agrees them he must send a draft VO to the engineer to cover them.

If he thinks an extra is invalid, or the rate or quantity is wrong, he should discuss this with the contractor, and may decide to delete the item, accept it, or substitute a rate or quantity he considers fair. He draws the engineer's attention to his decision when forwarding the contractor's payment application to him. If the contractor continues to dispute the item after the engineer's decision on it, the matter becomes a claim, is given a number and put on the Claims Pending file for further consideration.

It is preferable for each VO to cover one instruction only. However, one VO can be made to cover a series of related instructions, provided the instructions covered are itemized. A typical style of VO is shown in Fig. 13.5.

Bill Section 4: Inlet Works

Item no.	Item description	As bill		As measured			Bill rate	Amount
		Quant	Unit	Last cert	Since last cert	Total to date	£	£
	CONCRETE ANCILLARIES Formwork: Type B as specification Clause 411							
G213	Plane horizontal to suspended floor slab soffit. Width 0.2–0.4 m	3	m^2	3	–	3	34.12	102.36
G214.1	Plane horizontal to suspended floor slab soffit. Width 0.4–1.22 m	136	m^2	52	85	137	21.54	2950.98
G214.2	Plane horizontal to service ducts. Width 0.4–1.22 m	8	m^2	–	–	–	34.12	–
G215	Plane horizontal to suspended floor slab soffit. Width exceeding 1.22 m	152	m^2	100	52	152	22.55	3427.60
G223	Plane sloping to corbel at top of external wall. Width 0.2–0.4 m	43	m^2	20	23	43	33.38	1435.34
G224	Plane sloping to grit sump. Width 0.4–1.22 m	3	m^2	–	–	–	37.20	–
G242.1	Plane vertical to suspended floor slab. Width 0.1–0.2 m	198	m^2	120	78	198	12.45	2465.10
G242.2	Plane vertical to drainage channels rebate. Width 0.1–0.2 m	125	m^2	30	95	125	12.45	1556.25
				Total page 4/7 to summary				11 937.63

Fig. 13.4. Example of a contractor's form of application for payment. *Note*: The letter and first three figures of the Item No. are the CESMM Class and 1st, 2nd and 3rd Divisions respectively.

13.8 Authorization of dayworks

Some extra work ordered may be paid for at dayworks rates. This is adopted when no unit rates seem applicable, or when the amount of work required is indeterminate. A typical application of dayworks rates would be for offloading and stacking pipes delivered by the employer for use on the job – if no bill item for this has been allowed. In the UK the *Schedules of Dayworks Carried out Incidental to Contract Work* issued by the Civil Engineering Contractors Association (CECA) is widely used as a basis for charging dayworks. The most recent Schedule (issued in November 2002) provides for payment as follows:

- The cost of labour per hour is to be taken as the wages paid to a worker (inclusive of overtime, bonus, travelling time and lost time due to weather)

VARIATION ORDER
No..............
Job ....
Contract No. . ... Description ....
Contractor ....
In accordance with and subject to the Conditions of Contract you are hereby instructed to execute the following work:

The prices to be allowed for the above work shall be:

This work is additional to / substituted for work hitherto included in the Contract.

You are instructed to omit items of work as follows:

Drafted .... Signed ....
Resident Engineer Engineer

Date .... Date ....

ESTIMATED NET EFFECT ON THE COST OF WORKS

This Variation Order	....	increase/decrease
Add total effect of previous Variation Orders issued	....	increase/decrease
Total estimated effect	________	increase/decrease

Fig. 13.5. Layout of a variation or change order

plus 148 per cent; plus subsistence or lodging allowance and travelling allowances plus 12.5 per cent.
- The cost of labour sub-contractors and hired drivers is at invoiced costs plus 88 per cent.
- The cost of materials is the invoiced cost to the contractor of those materials plus 12.5 per cent.
- The cost of plant used is to be taken at the rates listed in the Schedule, and hired plant at invoiced cost (excluding driver) plus 12.5 per cent.
- Cash discounts up to 2.5 per cent are not deducted.

The percentage additions may vary from time to time as new editions of the Schedule are published following agreement reached in the industry. The plant rates in the Schedule are also reviewed regularly and cover most types of plant inclusive of fuel but exclusive of driver. For most items hourly rates are quoted, for others daily or weekly rates.

As an alternative to the CECA dayworks rates, tenderers may be required to quote – at the time of tendering – their rates against a list of labour categories put in a schedule attached to the bill of quantities. This is necessary for work done outside UK where no locally recognized schedule of dayworks rates may apply. The method is also useful for construction contracts in UK since the labour categories listed can be grouped into four or five classes according to skill or range of pay, and prices entered may be specified as inclusive of all oncosts and overheads. This simplifies the work of costing daywork sheets. Dayworks rates for plant may also be inserted by the contractor.

As soon as any dayworks has been authorized, the RE must inform the inspector or section engineer concerned, so that they can note the labour, materials and plant used on the operation. The contractor's foreman in charge of such dayworks will normally submit daily time and materials sheets to the inspector, for him to check and sign that they are correct. From these sheets the contractor makes up the dayworks account – typically as Fig. 13.6 – in duplicate and submits invoices to support the prices for materials. After checking and signing by the RE, one copy of the account is returned to the contractor for inclusion in the next monthly application.

13.9 Filing system for dayworks sheets

On a large job there may be a thousand or more dayworks sheets. It is, therefore, essential to set up a filing system to handle them. The following files will be necessary:

- DW1 – Dayworks sheets: New/To be dealt with.
- DW2 – Dayworks sheets: Checked/Pending signature.
- DW3 – Dayworks sheets: Signed and Returned to Contractor.
- DW4 – Dayworks sheets: Rejected and Returned to Contractor.
- DW5 – Resubmitted Dayworks sheets: Pending.

Arrangements should be made with the contractor for all dayworks sheets to be numbered consecutively, and all sheets must be submitted in duplicate. An exact copy of every sheet returned to the contractor must be kept, showing corrections and comments made on the sheet. Such sheets must be signed and dated by the RE. Care should be taken to ensure that sheets are filed as soon as they come in so none are lost. Comments can be written on a sheet returned; or on a signed note stapled to it, provided a copy is kept (one reason for a copying machine on site). It is too time-consuming to write letters to the contractor about dayworks sheets.

If the contractor maintains a dispute over a dayworks sheet after the RE has given his final decision on it, the contractor must be told to treat it as a claim and give it a claim number.

The file DW5 – for Resubmitted sheets: Pending – is for sheets returned by the contractor because he disputes a correction or rejection made by the RE. These sheets eventually end in Files DW3 or DW4 after being dealt with a second time by the RE. Alternatively the RE may, in such a case, send a letter to the contractor stating why his previous decision stands.

The problem of handling dayworks sheets submitted 'For record purposes only'– called 'FRPO sheets' – is discussed in Section 17.7. The filing of these will depend on the policy adopted with respect to them by the RE after consultation

DAYWORK/RECORDED COSTS

Sheet No :- 57

Contract No. TF/04

Description :- Excavate for & cut hole in 250 mm R.C. wall for 400 mm pipe

W.E. :- 23/6/01

Eng Ord :-

Basement extension

Name	Trade	M	T	W	T	F	S	S	O/T	T/T	Total	Rate	Bonus	£	Subsistence	£
Hunt A	Ch. hand				8	5					13	7.38		95.94		
Jaggers J	Lab				8	5					13	5.81		75.53		
Smith A	Welder				2						2	7.30		14.60		
Johns KA	Carp				3						3	7.30		21.90		
Casey J	Driver				4						4	7.08		28.32		
														236.29		
													+ 148 %	349.71		
														586.00		586.00

Plant	M	T	W	T	F	S	S	Tot	Rate	£
Poclain				4				4	29.30	117.20
Compres.				4				4	17.77	71.08
										188.28
										188.28

Materials			Rate	£	£
4m 50 x 75 sft wd	4	m	1.60	6.40	
2 x $1\frac{1}{2}$m chipbrd	3	m^2	3.45	10.35	
Cement	25	kg	£85/t	2.13	
agg. 150 kg	.15	t	8.40	1.26	
				20.14	
			+ $12\frac{1}{2}$ %	2.52	
				22.66	22.66
					188.28
				Total £	796.94

Certified Correct. Site Agent :- D.J.M Resident Engineer :-

Fig. 13.6. Dayworks sheet

with the engineer. If the decision is not to return them to the contractor, only one file for them is needed, the RE's factual findings and comments thereon being filed with each sheet. If some FRPO sheets are returned to the contractor after checking, two files will be necessary – FRPO/1 for incoming sheets; and FRPO/2 for copies of sheets returned.

13.10 Check of materials on site

Most contracts permit the engineer to certify payment on account of materials delivered to site but not yet incorporated in the works. Section 16.5 sets out the matters to be taken into consideration when assessing what payment can be allowed. Usually a contractor will only ask for payment on account of relatively expensive items, such as steelwork, reinforcement, pipes and valves, etc. Before agreeing to payment on account the RE will need to inspect the materials to ensure they conform to specification, and may need to get confirmation the contractor has paid for them or otherwise has ownership of them.

13.11 Price increase records

As mentioned in Sections 3.2(a) and 16.8 contracts extending over a lengthy period of time may incorporate a price variation clause under which the contractor is entitled to receive reimbursement of extra costs caused to him by inflation of prices for labour, materials and plant since the date of his tender. The amount due to him under this clause can be calculated according to a formula incorporated in the contract conditions, or by direct examination of the contractor's wages sheets and invoices received by him for materials or hire of plant.

If a formula is used, this will be re-calculated each month; hence a file of the price indices used for such a formula is necessary. However, if – more rarely – the price increase has to be calculated by reference to the contractor's wages sheets and invoices, a separate filing system for the extensive calculations involved will be necessary.

A file of basic costs at time of tender will be needed; another for wage increase calculations, and another for materials and hired plant. When invoices showing price increases are submitted, a check needs to be made to ensure that the invoiced quantity of materials shown on the invoice has been used on the job. For materials such as cement, aggregates, reinforcement, or fuel (for plant), a running total of quantities delivered as shown on the invoices must be kept to ensure the total does not exceed the possible use on the job. All invoices must be marked 'Seen' and initialled before return to the contractor, who must be instructed to file and keep them in case the employer's auditors wish to check the assessment of the price increases certified.

13.12 Supply contract records

On some projects separate contracts are let for the supply of pipes and valves, and other types of material to be incorporated in the works by the civil engineering contractor. Some of the materials, especially large pipes and valves, may be on long delivery so contracts for their supply have to be let before the construction contract commences. When this occurs it is essential to set up a stock-book in which a record is kept of all items ordered, and those delivered.

A typical stockbook page is shown in Fig. 13.7. It lists items ordered, where they are to be used, items delivered and where stored, and how finally used. The last three columns are useful to record a number of matters which, if not recorded, can cause confusion, such as pipes cut and the unused portion returned to stock, or a bend taken out but not used and returned to stock, etc.

For the financial book-keeping a Pipe (or Valve, etc.) Delivery Schedule should be kept in the style shown in Fig. 13.8. Under the columns headed 'Deliveries' the delivery position at any time can be known, and the tonnage weights entered can be used to calculate payments to the contractor for haulage on a tonnage rate basis. Under 'Payments' the checked invoice prices are inserted, and the date when the invoice is included in a certificate for payment. A transmission letter should always accompany transfer of invoices to the engineer, listing them by their reference and invoiced price. This acts as a check if an invoice goes astray.

The items when delivered would be stored in stockyards, from whence they need to be issued and accounted for in various parts of the work. Even if pipes are strung along the route of the pipeline, their location needs to be logged. Factors needing to be taken into account when setting up a system may be the following:

- Items need to be inspected for damage as they are offloaded.
- Some items may be delivered before the main contract starts, and some after.
- Some items may be supplied by the employer from his own stocks, but have to be collected by the contractor.
- Some further items may have to be ordered and delivered.
- Jointing materials, bolts and other small items will need storage under cover.

The RE should supply the contractor with a list showing where all items delivered are stored; and keep the list updated as more materials come in. He must make arrangements with the agent as to how materials are to be taken from stock. Usually the RE's pipeline inspector will take charge, he will tell the pipeline foreman where the appropriate pipes and specials are and will see that the right ones are taken.

If no proper stock-book is kept, there may be considerable wastage due to failure to make economic use of pipes and specials; or delay and extra cost can be caused by failure to use specials in the right place, so more have to be ordered.

PIPE STOCK BOOK

(1)	(2)	(3)	(4)	(5)		(6)	(7)	(8)	(9)
Item no.	Description	No.	Where required	Deliveries		Where Placed	Used	Surplus	Disposal
				Date	No.				
1	*750 mm $\times$ 4 m S. & S.*	*24*	*Pipeline 24*	*On site*	*8*	*Verge by crossing 8*			
				12 Sept	*8*	*Verge near entrance 8*	*15 & 1 No. 2 m s.p.*	*2 m p.p.*	*Scrap*
				1 Oct	*8*	*Entrance 8*	*6*	*2 No. (1 cracked spigot)*	*3 m length in tank: 1 No. stock*
2	*750 mm $\times$ 3 m S. & S.*	*1*	*Tank*	*17 Sept*	*1*	*Site Dump 1*	*1 No.*		
3	*525 mm 45° S. & S.*	*6*	*Tank inlets 2*	*1 Oct*	*6*	*Site Dump 4*	*2 inlets*	—	—
			outlets 2				*2 outlets*		
			Pipeline (crossing) 2			*Crossing 2*	*Not used*	*2 No. — (This column recorded in pencil and brought to date before stock check takes place)*	*2 No. stock*

Fig. 13.7. A page of a pipe stock book

PIPE DELIVERY SCHEDULE

Bill item	Description	No	Deliveries				Payments		
			Advice note	No.	Weight* (tonne)	Delivered	Invoice no.	Amount £	Passed for payment
D18	*400 × 150 fl. Tee*	3	*44/5838*	3	*0.30*	*15 Oct*	*5838*	*1 125.00*	*21 Oct*
							(Overcharge £46.0 deleted. See invoice.)		
D19	*300 × 150 -do-*	5	*14/7953*	2	*0.14*	*29 Oct*	*7953* }	*895.00*	*15 Nov*
			8/2572	3	*0.20*	*6 Nov*	*2572* }		*-do-*
D20	*150 × 5.5 m Sp/So.*	4	*14/7953*	4	*0.64*	*29 Oct*	*7953*	*250.00*	*-do-*
D21	*300 $22\frac{1}{2}$° bend*	2							
D22	*300 × 2.0 m fl/plain*	2	*14/7953*	2	*0.32*	*29 Oct*	*7953*	*454.00*	*15 Nov*
D23	*400–300 conc. taper*	1							

*The weight is only required if the contractor is paid by weight of pipes offloaded and hauled.

Fig. 13.8. A page of a pipe delivery book

The question can be asked – need the RE keep such a stock register if the contractor supplies the pipes and valves, etc.? The answer depends on the method of payment to the contractor. If he is paid unit rates for 'supply, lay and joint...' pipes and valves, then it is not necessary for the RE to keep a stock record – but the contractor will be wise to do so for the reasons given above. If, however, the contractor is to obtain the pipes and valves from nominated sub-contractors whose charges are reimbursed to the contractor, then the RE should set up the stock-book to check that mismanagement of items and unnecessary wastage does not occur.

Any materials left over on completion of the contract remain the property of the employer if supplied by him. This is another reason why control via a stock-book should be exercised by the RE, so that the employer does not get returned to him a miscellany of cut pipes of little use to him, but as many whole pipes and undamaged specials as possible.

13.13 Registers of test results

Test results on materials should normally be recorded on special forms to a format supplied by the engineer. A file for each type of test should be kept on site, copies of the tests being sent to the engineer. A general classification of tests for filing would be as under.

- Borehole logs, trial pit results, etc.
- Foundation material tests: grading curves; sample tests; analyses, etc.
- Earthwork tests: Proctor compaction tests; in-situ density tests; etc.
- Concrete tests: aggregate gradings and tests; cement tests; cube and beam tests, etc.
- Pipeline tests.
- Miscellaneous tests.
- Other manufacturer's tests.

Files should be fronted by a register of all tests taken. The particulars on the register must show where the sample is taken from, the date taken, date tested, and nature of test. Reference numbers for all samples must be given, and indelibly written on the sample packaging. Simple errors in labelling concrete test cubes, for instance, can lead to time-consuming, expensive and unnecessary alarms.

The position of all foundation or earthwork investigations, inspections, probes, samples, etc. should be marked on a plan. It is essential to keep a second up-dated copy of this plan since loss of it can greatly reduce the value of such investigations.

On many civil engineering projects equipment installed of various kinds may need to undergo performance tests, some of which may be extensive lasting several weeks. Also logs of various observations of the performance of the works may be needed, such as movements of a dam during filling or settlement of earth structures. There may also be test results on mechanical and electrical plant, crane test certificates, logs of underdrain flows and pore water pressures, settlement readings, and many other matters. The results of these tests and observations must be collated and preserved, since they will all need to be supplied to the employer when he takes over the project. Some test observations may have to be started as soon as part of the works are constructed and kept going for the rest of the job. All these matters are important because all such performance records may need to be summarized in a suitable form and presented to the employer in a Completion Report.

13.14 Photographs

Photographic records of the project can be invaluable and their cost is small relative to their worth. The following list shows the type of photographs that can prove useful:

- Photographs before any work is undertaken:
  - of the site generally (e.g. picture views, etc.);
  - of any buildings to be demolished;
  - of the condition of any adjacent buildings liable to be affected by the works;
  - views of access tracks and public roads to be used by contractor, plus close-up photos of surfaces of public roads before use.
- Monthly progress photographs of the work during construction.
- Photos of technical matters (such as the nature of foundation material, etc. covered up) which need recording.
- Photos to illustrate any problem that has occurred on site and which needs to be reported to the engineer for comment or advice.
- Photos of the completed works; particularly after all rubbish has been cleared away.

It is essential that all prints are marked on the back indicating the job, the feature shown, date taken, and the negative reference. Filing of photos is not easy: the classification above is a starting point. Over-large albums should be avoided since they often do not fit in standard shelving: ultimate box-file storage is usually most practicable.

13.15 Record drawings

The engineer's agreement with the employer will usually require him to provide the employer with 'as built' record drawings of the completed structure. Normal practice is for the RE or his staff to mark all amendments or additions in red on a copy of the contract drawings, the original master copies of the contract drawings then being amended. This is not entirely satisfactory because not all contract drawings are relevant or sufficient for record purposes. To make a good set of record drawings may involve discarding a number of contract drawings; and using 'cut-and-paste' methods to make up a single drawing from parts of contract drawings, or producing completely new drawings. Foundation drawings which have been prepared on site to show the contractor precise dimensions and levels for foundation excavations should be included among the record drawings.

A drawing showing important details of construction can often be made up from copies of sketches supplied to the contractor. Such details can be invaluable in tracking down the possible cause of some after-trouble, such as damp penetration.

In general record drawings should give:

- a good detailed layout plan of the project;
- a detailed foundation plan;
- floor plans for inside of buildings;
- plans showing the location of everything underground and what depth it is;
- details of construction where these are hidden from view.

Where new drawings produced on site are ultimately required in a digital format, this may have to be carried out at the engineer's head office.

It is not necessary to show all the minutiae of construction which can be seen or measured on site after construction is completed. Copies of reinforcement drawings are usually supplied separately bound from the record drawings. Their main purpose is to show what size and spacing of reinforcement was used. They cannot show the exact position of bars.

On clearing up the site supervision organization, all drawings superseded and not applicable to the finished works should be destroyed. This is important, because if any drawing remains of some proposal not adopted, confusion may later be raised as to how a structure was actually built. This can give rise to serious difficulties when, for instance, later repairs have to be undertaken on a dam, or tunnelling below a structure has to be undertaken. The position of all services underground should also be marked to avoid trouble when additional services have to be laid later.

13.16 Other records

A **job completion** report may be of significant value, both for publicity purposes and for logging down experiences that can be of value later. The salient facts about the project should be listed – client; description of works; purpose, sizes and outputs; designers and contractors involved; dates started and finished; budgeted cost, final cost and chief reasons for any difference; date of opening ceremony, etc. A short report should be attached of any significant technical problems encountered and how they were overcome. The report should concentrate on such matters that, from experience, can form useful guidance for future designers and those who draw up contracts.

The RE will save himself much later time and trouble if, as soon as any equipment arrives, he takes charge of the drawings and instruction manuals for it and asks the manufacturer for two more copies of them or gets them copied locally. They are of importance to the employer and should be collated in some orderly fashion.

A file should be made listing the names and addresses of all equipment suppliers, and of the suppliers of key materials used in the works, such as ceramic tiles, facing bricks, cladding, etc. The file should give details of what was supplied and the date it was ordered. A copy of this file should be given to the employer for whom it will be valuable when it is necessary to repair or replace items, or if performance problems occur. Instruction manuals and plant test data, such as performance curves of pumps, turbines and motors, should all be collected, and two sets of each should be obtained to supply to the employer.

An essential requirement for works in the UK is production of a **Health and Safety File** (see Section 10.4) at completion of the work. This file is to be handed over to the client to ensure that information is available to him on any hazards which may affect anyone doing maintenance work or future construction work. Under the CDM regulations the planning supervisor is responsible for ensuring that the file is produced but input will be required from all concerned and it may fall to the RE to ensure that the various contractors involved produce relevant information at completion of the job.

14

Programme and progress charts

14.1 Responsibilities for programming the construction

The contractor is responsible for producing a programme for construction for the job, though he must comply with any special requirements laid down in the contract documents. Under the ICE conditions the contractor must submit his proposed programme within 21 days of being awarded the contract (Clause 14(1)). Within a further period of 21 days (see Section 9.5 Item 4), the engineer must accept or reject it, or call for more information on it; if not, he is deemed to have accepted it (Clause 14(2)). If the engineer calls for more information, the same time limits are repeated.

The programme for construction may therefore have been agreed before the resident engineer goes to site. But the resident engineer needs to check what it requires with respect to (1) the provision of further drawings and information to the contractor, and (2) the provision of any materials or services to be supplied by the employer under separate contracts he has entered into, or which are to be obtained by the contractor from nominated suppliers.

The delivery times for nominated suppliers and sub-contractors should have been quoted in the main civil engineering contract, requiring the contractor to allow for them when drawing up his programme. It is prudent to add 'margins of safety' to the delivery times quoted in the contract because (a) the contractor can only place orders with them after he is awarded the contract, and (b) the nominated firms might not deliver on their promised time, causing a delay to the contractor's programme enabling him to claim for delay.

Hence the resident engineer should check the current delivery times quoted by nominated firms and advise the contractor of the latest times he must place orders.

14.2 Difficulties with nominated sub-contractors or suppliers

The use of nominated sub-contractors or nominated suppliers can cause many problems because the engineer cannot interfere in the terms of the sub-contract (see Section 15.8). The sub-contractor or supplier may refuse the sub-contract because of disagreement with the contractor on liability for damages (see Section 7.8), trade discount and terms of payment, or some extra charge the contractor wants to make for services he provides. Sometimes a sub-contractor refuses to accept an order from the contractor for reasons he will not disclose – usually due to some past experience with the contractor. Although some problems can be overcome by careful detailing of all necessary provisions in the specification, there is never any certainty the sub-contract will be signed. If not signed the purpose of nomination is frustrated, and re-nomination may be necessary causing the programme to be disrupted.

To avoid the problems of nomination the specification can specify the work required and leave the choice of sub-contractor to the contractor; with the proviso that the sub-contractor must be approved by the engineer. Nominated suppliers can also be avoided by specifying items, where possible – 'As Messrs. XYZ's product or similar'– leaving the onus on the contractor to choose his source of supplier.

But this is not always possible when, for instance, the employer wishes to use facing bricks available from only one supplier. An alternative then is for the employer to place the order for the bricks direct with the supplier, making arrangements for their offloading and stacking; with the engineer denoting in the contract for construction where the contractor is to find such bricks, etc. The resident engineer needs to ensure all such arrangements are being made in due time to avoid delay to the contractor.

Fortunately many small items to be supplied by nominated suppliers are not crucial to the contractor's programme, or not required until the finishing stages of the contract. These usually do not give rise to many problems, and the contractor can be encouraged to order them in good time by the engineer certifying part payment of their value under 'materials on site'.

14.3 The role of the resident engineer

The resident engineer should contact separate and nominated sub-contractors or suppliers, advising them construction has started, getting them to confirm their delivery times. He should also make sure that all technical queries are settled. Where suppliers have to manufacture substantial equipment, he will check their progress, and may visit them to make personal contact. He will do everything possible to prevent any delays occurring and, if he sees some delay is unavoidable, he will inform the engineer and make suggestions as to how any consequential delay to the contractor can be avoided. However he can only make direct contact with nominated suppliers or sub-contractors before the contractor places his order with them. After the contractor has placed his order, any contact with a nominated sub-contractor or supplier must be via the contractor, unless the contractor permits otherwise.

There may be other matters with respect to the programme the resident engineer should look into. In some cases the employer may require access through the project area for his other works. Or perhaps work by the contractor must necessarily interrupt services which the employer relies upon, such as electricity, drainage, water lines, etc. There may therefore be a strictly limited time which the employer can tolerate such interruption; and he may prefer the interruption to occur at some particular time of year rather than another.

The influence of the weather may be an important factor to take into account when examining a contractor's programme, especially if the contract involves substantial earthwork construction. The resident engineer may need to discuss with the contractor where he thinks the programme should include optional strategies according to weather. He should be able to advise what sort of measures could be taken to minimize the effect of weather.

The resident engineer has to appreciate that a contractor must ensure his programme for construction fosters efficient, economic working. Once he has brought men and machines onto the site he will want to use them continuously until their tasks are completed. Also he will want their output to be as near as possible to their maximum. Hence the resident engineer must appreciate that, on occasion, a contractor has to 'make do' with what plant and men he has on site, because the expense of bringing in more to do a 'one off' job is too great to be economic. The resident engineer can only interfere when he is certain that some method proposed by the contractor will result in unsatisfactory work or some unacceptable risk to safety.

14.4 Watching and recording progress

From the agreed programme it is useful for the resident engineer to draw up a list of dates by which different operations must be undertaken as shown in Fig. 14.1. If there are several contracts let for the construction of a project, the list will be essential for co-ordination of the work of different contractors. It is useful as an overall guide for checking if the contractor is keeping to time and as a reminder what future actions need to be taken.

Figure 14.2 shows a typical bar chart for a single structure. The length of the broad bands show the time duration expected for each operation; these are coloured or hatched in as work proceeds to show how much of an item of work has been completed. The solid black lines indicate the actual time periods taken to achieve the quantity entered in the broad band. Figures can be written in to show the quantity of work done by the end of each week, as compared

Time Chart

Contract Times				
		Tank	*Admin. Block*	*Pumping Station*
1st Year	Jan.			
				Order windows,
	Feb.		Order roof trusses	roof beams
	Mar.	Excav.	Prelim.-heating	Founds. in
	Apr.			
	May			Walls
	June	Walls		
	July			Floors
	Aug.		Order slates	
	Sept.		Order guttering	Pump founds.
	Oct.			
	Nov.			Piping
	Dec.			
		Cols	Roof on	Windows
2nd Year	Jan.			Roof beams
	Feb.	Floor		Roofing
	Mar.			
	Apr.	Main		Roof on
		laying	Plumber	
	May	Roof		Commence install.
	June		Heating	machinery
	July	Asphalting	Glazing	Doors. Glazing
	Aug.		Electrician	Electrician
	Sept.	Testing	Plastering	Plastering
	Oct.	Finish embanking	Seeding	
				End install.
	Nov.	Seeding	Decorating	machinery
	Dec.	Manholes		Testing
3rd Year	Jan.			Painting
	Feb.		Final decorating	Roads – final
	Mar.			
	Apr.			
	May		Completion	

Fig. 14.1. A list of target times from which a bar chart can be derived

with the quantity planned to be done. Usually the planned output is shown in black, and the actual performance is entered in red.

A bar chart is easy to interpret, and keep updated. An agent will almost always have a bar chart pinned up in his office which his site engineer updates each day. The resident engineer will have one of his own, made up from the monthly measurements. The disadvantage of a bar chart is that it is difficult to apply to a complex project. The bar chart shown in Fig. 14.2 can only be meaningful in relation to one structure. If several structures are to be constructed under a project, it is not satisfactory to lump all their excavation,

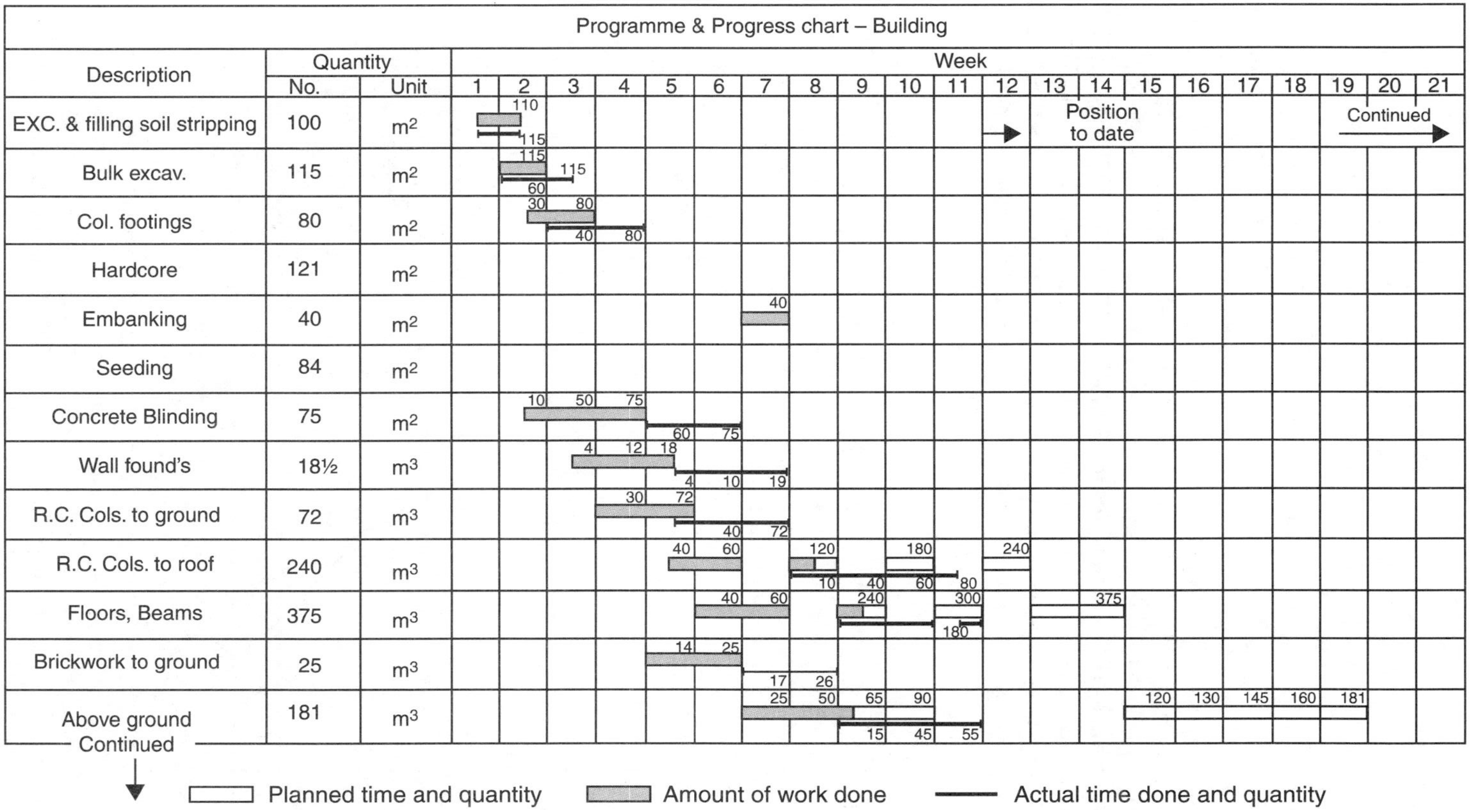

Programme & Progress chart – Building

Description	Quantity No.	Quantity Unit	Wk 1	2	3	4	5	6	7	8	9	10	11	12	13	14	15	16	17	18	19	20	21
EXC. & filling soil stripping	100	m^2		110; 115										→	Position to date						Continued →		
Bulk excav.	115	m^2		115; 60	115																		
Col. footings	80	m^2		30	80; 40	80																	
Hardcore	121	m^2																					
Embanking	40	m^2							40														
Seeding	84	m^2																					
Concrete Blinding	75	m^2		10	50	75	60	75															
Wall found's	18½	m^3			4	12	18; 4	10	19														
R.C. Cols. to ground	72	m^3				30	72	40	72														
R.C. Cols. to roof	240	m^3					40	60		120; 10	40	180; 60	80	240									
Floors, Beams	375	m^3						40	60		240		300; 180			375							
Brickwork to ground	25	m^3					14	25	17	26													
Above ground Continued ↓	181	m^3							25	50	65; 15	90; 45	55				120	130	145	160	181		

Key: ▭ Planned time and quantity; ▬ (shaded) Amount of work done; — Actual time done and quantity

Fig. 14.2. Part of a bar chart for constructing a building

concreting, brickwork quantities etc. together. However, if there are only a few structures to deal with, bar charts for each will be practicable.

An alternative is to use a computer to display bar charts at several levels of detail. For resource planning and material ordering purposes the detailed operations required for each structure can be shown on one chart, and these can be summarized into one bar on another bar chart, which in turn can be used to represent progress on the project as a whole. Critical linkages between operations can be fed into the program as for network diagrams (see Section 14.5), and the resulting critical paths and 'floats' can be derived. Adjustment to a detailed bar chart – perhaps due to some delay or extra work – is automatically reflected in the overall summary bar chart display. Of course the operation of such software requires investment in the time and the use of skilled operators, so the cost may only be justified for a large complicated project.

Another form of progress chart is to mark up the cumulative value of work done on a graph as illustrated in Fig. 14.3. However, to ensure that such a chart

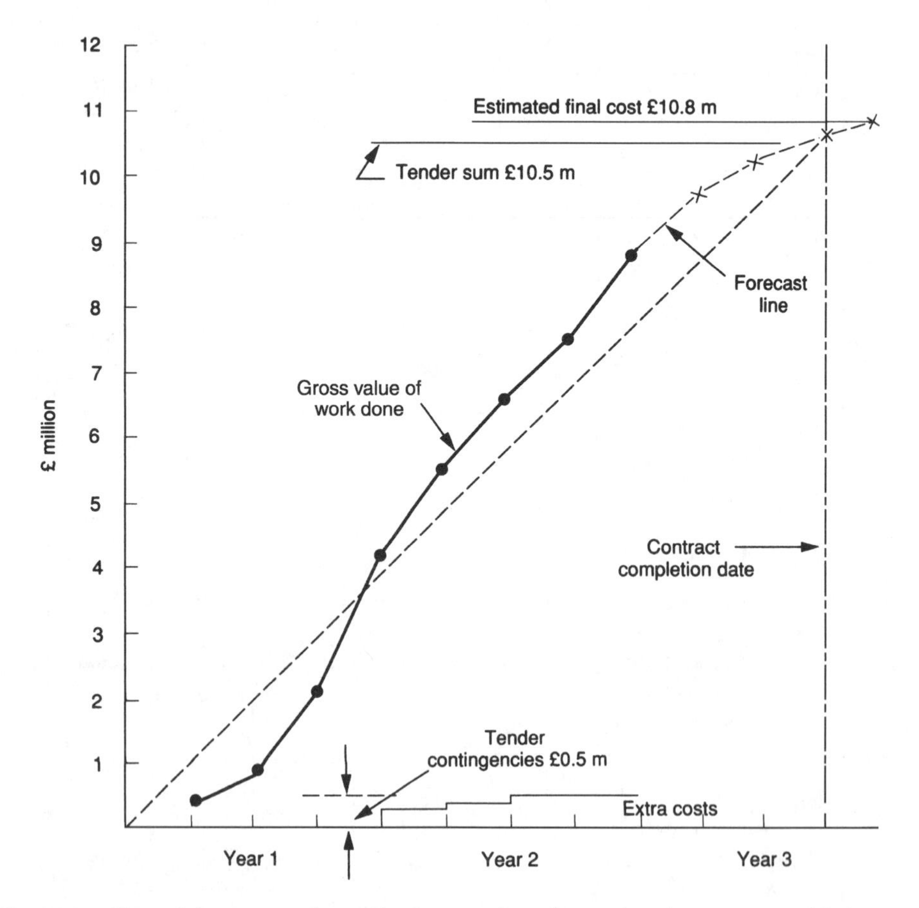

Fig. 14.3. Financial progress chart. The forecast line shows that the contract is likely to be completed about 3 months late at £0.3 million excess cost

does not give a false impression of progress the following adjustments are advisable:

- The 'target' cost should include the contingency money provided. A lower line at the base of the graph can show this contingency money and the expenditure against it, thus revealing where any exceedance of the target is due to excess contingency money spent.
- If there is a major item of expenditure incurred late in the contract (such as for supply and installation of plant), the value of this should also be excluded, or be plotted separately as an additional line.
- The cumulative valuations plotted monthly should not include retention money deducted, nor payment on account for materials on site, nor reimbursement of increased costs of wages and prices.

Although such a progress chart is not exact, it is a good indicator of progress. Usually the plot of valuations forms an S-curve, having its steepest inclination during the central part of the contract period when productivity should be at its greatest. As a consequence, if the plotted line of valuations does not rise above the straight line sometime during the middle period of the contract, then almost certainly the contract will finish late.

One advantage of a progress chart of this kind is that, when an employer needs an estimate of future rates of expenditure, this can be estimated by sketching in an S-curve of the type shown and reading off the monthly rates of expenditure it implies, less any retention money held back.

Some standard conditions of contract, such as those of FIDIC (see Section 4.3) require the contractor to produce a cash-flow forecast along with his programme for construction. It must be remembered, however, that a cash-flow from interim payments does not necessarily represent construction progress, because deductions are made for retention and additions may be made for advance payments (if any), and materials delivered to site.

Similar types of cumulative output chart can be applied to specific types of work. One for concrete work to different structures is shown in Fig. 14.4. The horizontal bands show to scale the amount of concrete to be placed in each structure. The cumulative amount of concrete placed month by month can be plotted for each structure, and their total shown separately. Such a chart is useful for indicating where slow progress is occurring.

For a mainlaying contract the type of progress chart shown in Fig. 14.5 can be used. It is self-explanatory. The profile of the main may have to be shown in a condensed form; it also shows where specials are required.

14.5 Network diagrams and critical path planning

A network 'diagram' has been referred to in the preceding section. It lists each activity required to complete a project giving each a reference and estimated duration – usually in weeks. An assumption is made concerning the order in

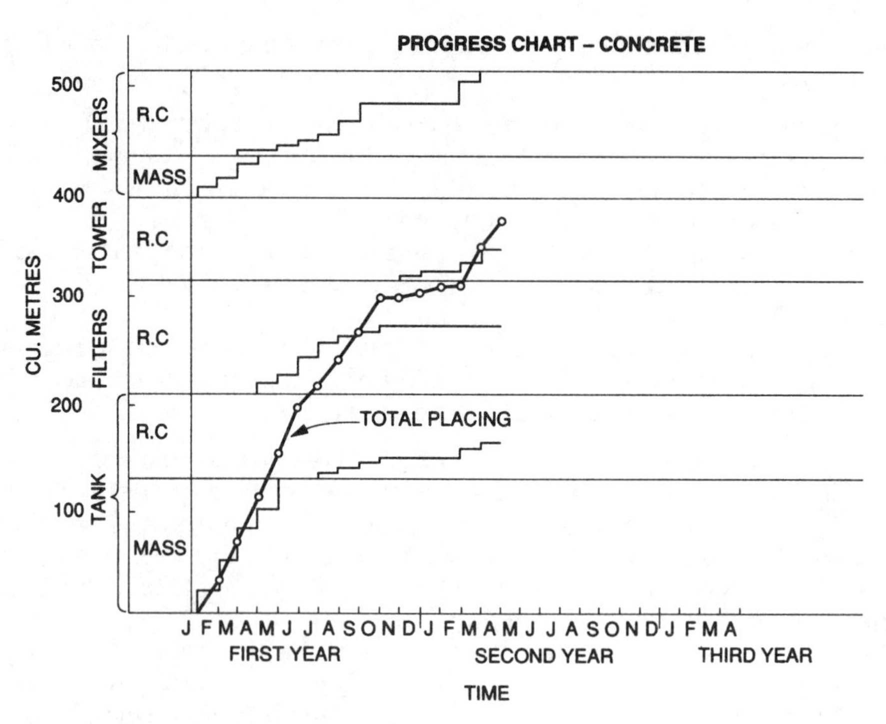

Fig. 14.4. A progress chart for concrete work

which construction will proceed. A network of 'connections' are then made between activities, stipulating the earliest and latest times each can start relative to some prior activity. The 'diagram' thus comprises many parallel strands of activities interconnected at many points where an activity cannot be started before another activity is wholly or partly complete. The computer traces through this network to find the longest total time taken by some unavoidable sequence of activities. This is the **critical path** which determines the minimum time to complete the whole project, on the assumed order of doing the work.

Modern computer network programs reproduce the analysis findings as bar charts on the screen, with differing colours for critical and non-critical activities, and showing **float** times also – the latter being the spare time available for completion of a given activity before it becomes 'critical'. This presentation makes the results of the analysis easier to understand.

But a difficulty is that the computer program has to be continually updated to include changes in the order of construction and changes in duration times for activities, which can arise from many causes, such as weather, troubles with labour or plant, delays in getting materials, etc. Hence critical path planning is sometimes started, but then abandoned because the program has to be continuously updated and the critical paths revealed are of little practical value since they can alter with every update. Bar charts are preferred for most jobs, since they involve less work, are easy to draw and keep updated.

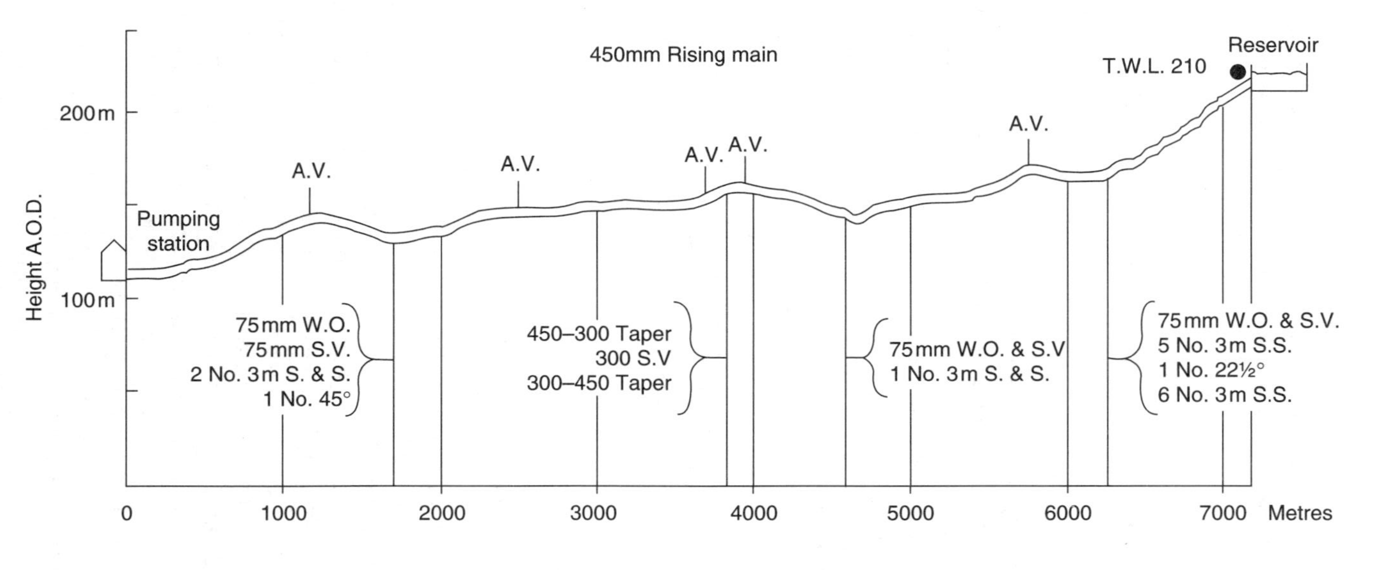

Pipes delivered	2.9.93	16.9.93	23.9.93	16.9.93
Laid	30.11.93	31.12.93	31.1.94	31.1.94
Tested	15.12.93			
Reinstated	14.1.94			

Fig. 14.5. Pipeline progress chart

Contractors sometimes use the critical path method to support a claim for delay; but the same problem applies that any critical path is based on only one particular order in which work is constructed, and other orders may be possible.

14.6 The part played by the agent in achieving progress

It is the contractor's agent who has on-the-spot responsibility for programming the work and keeping progress in line. The resident engineer's job is to assist the agent, if asked, and provide any information that the agent needs or that will be helpful to him. As the work proceeds the resident engineer will keep a check on progress, and must advise the engineer when unacceptably slow progress is occurring. Before acting formally in this matter the resident engineer should put his comments to the agent, seeking to find out why work is going slow and endeavouring to persuade him to take steps to speed up construction. He must also identify causes of delay for which the employer is responsible.

A contractor's slow progress can be caused by many factors – lack of labour, lack of skilled key men, a weak general foreman, or an agent not sufficiently decisive or good at organization, or tending to under-estimate the difficulty of a job and failing to foresee problems arising. Sometimes the cause may lie with the contractor's head office, such as slowness in getting materials or equipment to site. This may be indicative of the contractor being outstretched, either organizationally or financially. It is important that the resident engineer gets sufficient information to give the engineer reliable advice as to where the cause of slow progress lies because, if the lack of progress continues, the engineer will have to take up the matter formally with the contractor.

A good agent is an inestimable benefit to a project. He automatically thinks in terms of the 'critical path' that lies ahead, and has clearly in his mind where the job 'ought to be' in a month's or 2 months' time. But to get there he has to make many decisions in the present. He has to seize opportunities, overcome delays, take extra work into account, suffer inefficiencies of labour and breakdowns of plant, find solutions to unexpected problems, face the vagaries of the weather and, despite all these, keep the work going at the required pace to gain his targets. The immediate targets are short term – this week's in detail, next week's in outline. If he can achieve them, he knows they are within the longer term strategy he has already worked out.

He has also to be aware of the need to have safety margins of time in hand for overcoming all sorts of difficulties that his experience tells him will inevitably crop up, even though he cannot forecast the precise form they will take. Many factors influence his judgement. He will be quick to detect when things are in his favour – when weather seems to promise fine, when the spirit on the job is good and the men are working efficiently as a team – and, grasping such opportunities, he will use them to drive the job onwards, knowing that one success leads to another. By experience and force of his personality he may pull the job ahead of schedule and complete before the promised time.

14.7 Completion

One of the duties of the engineer is to decide when the works can be considered complete. Most contracts require that the works are substantially complete before the engineer can issue his certificate of substantial completion. Unless the contract has specific requirements to be met, substantial completion does not require that every last item is finished, but it is generally taken to mean that most of the works are done and the project can be put to safe and effective use by the employer. There will no doubt be some items outstanding and conditions, such as ICE (7th edition), allow for this while requiring that any outstanding work is completed as soon as practicable during the defects period or to an agreed timetable. Items left to be finished later would be those which do not affect operation, such as painting; or may even include minor structures, such as a gatehouse not essential to use of the works.

A shrewd contractor will be looking to get his certificate of completion as early as possible and may apply to the engineer as soon as he thinks he has any chance of it being allowed. Clause 48 of the ICE conditions (7th edition) allows for such an application which must, however, be accompanied by an undertaking to finish outstanding work. The engineer must then either issue a certificate, or state what needs to be done to complete. The resident engineer must advise the engineer of matters still to be completed and say if he considers the contractor's application for completion is too early. Before the engineer issues his certificate of completion he will need to check with the employer to ensure that he has staff available to take over completed works.

14.8 Estimating extension of time

Failure to complete in time may make the contractor liable to pay liquidated damages, as specified in the contract, to the employer. Hence at the time a completion certificate is issued, the engineer must decide whether any extension of the time for completion is allowable. Extensions of time may have to be allowed if the contractor is delayed by problems for which the employer is liable under the contract. These should have been notified as they arose and considered at the time (see Section 17.10). There may also be sections of the work which the contract requires to be completed earlier than for the project as a whole. It is important to note that not all delays will lead to an extension. Some may not affect the path to completion and others may be matters for which the contractor is responsible. It is necessary to check with the contract conditions (such as ICE 7th edition Clause 44) to identify which delays are to be taken into account, such as 'exceptional adverse weather conditions', etc.

Assessment of extensions of time may have to be made several times during a long contract and their cumulative effect will have to be assessed at or near completion. The assessment needs to take account of all known circumstances at the time, independently of what may happen later. Also the estimate is unlikely to be precise because of complicating factors, such as delay periods overlapping. The engineer has to decide whether delays fairly entitle the contractor to an extension, so it may be that reference only to the contractor's programme is insufficient, since other factors need to be taken into account to produce a fair result. To make this judgment the engineer needs experience of constructional processes and their limitations.

An early look at a delay with reference to the contractor's initial programme, should take into account that this is only a statement of intent and that the contractor is at liberty to change his programme. Some operations will, in the nature of events, take longer or shorter to complete than anticipated. A more realistic approach is to examine the as-built record of construction to see if – had the delay not occurred – the construction could have been completed faster, taking into account any changes made by the employer which have caused delay. The advantage of this approach is that it is based on actual performance including any mitigating measures that were taken to reduce delay.

A further discussion of assessing delay and evaluating any payment due is given in Sections 17.10 and 17.11.

14.9 Estimating probable final cost of works

When a construction project is moving towards completion the employer may ask for an estimate of its probable final cost. To deal with this it is a help if the bill of quantities for the project is divided into separate bills for separate structures. This makes it easier to identify where extra costs have arisen and where further extras can be expected. The following type of analysis may then be adopted:

- An analysis of amounts incurred under variation orders is made, dividing them out over the separate bills. The analysis should include VOs pending. If a VO covers many items spread over several bills, it can be allocated to general contingency money, to save time on too much detailed analysis.
- The page totals in the original priced bill of quantities should be compared with the latest page totals for interim payments certified to date. Additions should be made where, by examination, it is thought payment for items on that page will come to £500 or more than the original page total shown. Deductions will be made where it is expected a page total will reduce by £500 or more because of items omitted, or for which payment under a VO has been substituted. Smaller differences are ignored. The total for each bill for a separate structure is then derived by adding up the revised page totals.
- The probable final cost outcome of the main construction contract can then be assessed as shown in Table 14.1.

Table 14.1

Estimated Final Cost of Contract XYZ

	Bill 1 (£)	*Bill 2 (£)*	*...etc. (£)*	*General contingency (£)*	*Total all bills (£)*
(a) Total of original bill items	....	....	....	Excluded	....
(b) Estimated change to bill items	....	....	....	....	....
Revised bill items	....	....	....	....	....
(c) Payable under VOs issued and pending	....	....	....	....	....
(d) Allow for further VOs	....	....	....	....	....
Total measured items	....	....	....	....	....
(e) Dayworks:					
Paid to date				....	
Allow for sheets in hand not yet paid				....	
Allow for further Dayworks to come				....	
					....
(f) Claims paid and claims agreed but not yet paid					....
(g) Allow for					
Claims pending not yet agreed					....
Possible other matters					....
Estimated total final cost					£ ....

Notes: (b) includes estimated extra costs due to increased quantities, less deductions for items superseded by new items covered by VOs.
(c) is the total payable under VOs, divided over the bills as far as ascertainable, any unallocated balance being put under general contingencies.

15

Measurement and bills of quantities

15.1 Principles of pricing and payment

In the simplest contracts for construction, the amount to be paid to the contractor may have been pre-agreed as a lump sum, or a series of lump sums relating to different items to be provided, and payment will depend on these being completed. In principle many turnkey types of contract and some design and construct contracts follow similar arrangements although, in practice, things seldom turn out so simple.

For many construction contracts, however, the works cannot be precisely defined at the time when the contract for construction is entered into, such as the depth to which foundations should be taken or pipelines and sewers laid. Also there may be additions or alterations found necessary when conditions on site are not those expected. In such circumstances a re-measurement type of contract may be the most suitable. Such a contract must set out exactly what is to be measured for payment purposes and when the payments are to be made. Hence it also needs to show, either specifically or by implication, what work is included in the prices to be paid.

Under most standard forms of contract the contractor undertakes to carry out the works described in the specification and shown on a set of drawings included in the contract. This obligation of the contractor is one which he takes on independently of the terms set out in the contract as to how and what he will be paid for various items of work. When a bill of quantities is used for the basis of payment, the specification and drawings describe what is required in detail, and the tenderer has to consult these when he prices a bill item and take them into account. He has also to see from the rest of the contract what are the obligations he must cover when he enters a price against a bill item.

The bill of quantities is, of course, of importance in the tendering process because it allows a reasonably fair financial comparison of tenders, but it does not limit what is to be built, nor limit the contractor's obligations. Under the ICE and similar conditions the bill of quantities is merely used for the ultimate task of pricing the works actually constructed. What is to be built will depend on the drawings, specifications and instructions issued by the engineer which the contractor is bound to follow. What is to be measured as a basis of payment is fixed by the method of measurement set out in the contract.

15.2 Methods of measurement for bills of quantities

The items in a bill of quantities can either list the work to be done in great detail, or can use fewer items, many of which are 'inclusive'. A common example of an inclusive item used on pipelaying contracts is – 'Manholes Type A complete' – measured by number of, which means as Type A shown on the contract drawings including concrete walls and roof, step irons and iron access cover etc. all as specified. But for other types of work detailed listing of items might be required to allow for possible variations and adjustments to the work shown on the drawings. On a large project, a considerable amount of detail is inevitable, not only because many different types of work will be involved, but also different circumstances or locations for similar types of work will apply.

When items are inclusive a tenderer has to ensure he has allowed for all such subsidiary matters in his price. He takes a bigger risk with his price for inclusive items than if the various details were separately itemized because, if by error he omits to allow for some work included, his resulting underprice is multiplied if more of that item is ordered. Hence his rates may be high for an 'inclusive' item and this affects the employer also, since increased quantities of that item may result in a disproportionately increased payment by the employer.

The choice of method of measurement for civil engineering work in the UK lies between using the civil engineering standard method of measurement (CESMM) as described in the next section; or using some different method; or using the CESMM for some types of work and a different method for the rest of the work. Where the standard method is used it must be followed, and any departures from it must be clearly stated in the preamble to the bill or in item descriptions. If the standard method is not used at all, Clause 57 of the ICE Conditions must be amended and the method adopted must be clearly defined.

The standard method itemizes work in considerable detail and therefore reduces the risks to both parties when admeasurement of the work takes place. There are also computer programs devised to assist billing and pricing by the standard method and this may be of use to tenderers' estimators who are familiar with the method. But the CESMM is complex, producing much detailed itemization of the works. Hence it is common to adopt the standard method for some work, and a different method for other work to reduce the number of items required. This different method usually comprises items of an inclusive nature.

For overseas contracts CESMM is seldom followed. Instead methods may conform as much as possible with the local practice used by the local state or public authorities.

15.3 The ICE standard method of measurement

The ICE standard method of measurement is not mandatory, but the ICE conditions require the method to be used – 'unless general or detailed description of the work in the bill of quantities or any other statement shows clearly to the contrary'. The most recent (3rd) edition of the CESMM was published in 1991 with corrections in 1992, and is commonly referred to as **CESMM3**. The standard method is not a contract document, and thus is not used in interpreting the contract – except in so far as its provisions are repeated in the contract documents (see below). Its use is solely as a recommended method of measurement in conjunction with the ICE Conditions, and is generally on the basis that all the works will be designed by the employer or his engineer.

Problems in the use of the standard method. The use of CESMM over a number of years has indicated several potential problems in the compilation of bills of quantities and measurement. There are seven introductory sections printed in the method and, although these are largely guidance notes for people preparing a bill, some parts need to be included in a contract, such as the parts dealing with adjustment items or method-related items. Other parts of the guidance notes may need exclusion to prevent the parties trying to alter the method of measurement after award of contract.

The parts of the CESMM's preliminary sections which are needed should be written into the contract documents themselves. Also the measurement rules may not apply, or may not be suitable if the contractor is required to undertake some element of design, such as in providing bearing piles.

However, it is not usual to depart from the *units* of measurement in CESMM3, or the *measurement rules* and *coverage rules* set out in the Work Classification sections of the CESMM. The measurement rules say, for instance, that when measuring concrete volume there is to be no deduction for the volume occupied by reinforcement, rebates, grooves and holes up to a certain size, etc. The coverage rules denote, for instance, that an item for supply of timber components includes their fixing, boring, cutting and jointing. Such rules are useful in making clear what the bill items are intended to include.

The standard method results in lengthy bills and for some types of work may seem to give an unnecessary number of items, or to divide work down into such detail that considerable thought has to be given to billing and pricing. Modifications to the method must, however, be very clearly put in the contract in order to avoid the possibility of the parties trying to argue for re-measurement or additional measurement where this was not intended.

For instance, instead of itemizing painting of step irons, ladders, etc. separately, a sub-heading can be put at an appropriate position in the bill stating: 'The following items to include painting after fixing'. The CESMM mentions that a line must be drawn across the description column in the bill below the last item to which the sub-heading is to refer. If, however, there is so much painting to do that a contractor would probably sublet it to a painting subcontractor, a non-CESMM item might be put in the bill of quantities, such as

'Painting items N1–N13 after erection ... Lump Sum'. Alternatively a provisional sum for painting can be entered. By such procedures the number of items in a bill can be reduced.

Description of items. The standard method states that item descriptions are to avoid unnecessary length, their intention being to – 'identify the component of the works and not the tasks to be carried out by the contractor'. Nevertheless descriptions according to the CESMM method tend to be lengthy in some cases. Each item has a letter and three-figure code number which identifies the work required according to the CESMM classification; but the code descriptions are not taken as definitive and, to avoid ambiguity, the actual descriptions have to be written out in words.

The bill items have also to be read in conjunction with the specification and the drawings – and it is an essential matter for the drafter of the bills to ensure all these relate. The location of items may need to be specifically stated also, and any additional description rules specified by the CESMM must be followed, in order to ensure that all detail necessary to identify the work is given, as required by the method. This ensures that items can be priced properly and their application to the work on site identified easily.

15.4 Problems with classes of work and number of items

For most works of any size there should be separate bills for obviously separate parts of the project. This clarifies the location of work under bill items, makes it possible to cost structures separately, and may be needed if completion of certain parts of the work is required by some stated earlier time. Within each bill the items will be classified into different types of work, always taken in the same order in all bills.

The standard method lists 26 classes of work labelled A–Z; Class A being for general items (more commonly known as 'Preliminaries'); Class B is for site investigation including sampling and laboratory testing; Class C for geotechnical processes, such as grouting and construction of diaphragm walls; Class D for demolition and site clearance. Thereafter there follow classes for the common constructional operations – earthworks, concrete, pipework, etc. – through to Class Y which is for sewer and water main renovations. The final Class Z is for 'Simple building works incidental to civil engineering works' and covers carpentry and joinery, doors and windows, surface finishes and services, etc.

Not all the 24 classes of construction work, B–Y, will normally be used on most projects, and a problem is that if the project includes a large building, the items under Class Z may be numerous and so need sub-classification. There may also be some difficulty in deciding where to bill certain types of work to achieve a logical order, since some work which would normally be considered part of the finishing building trades, such as painting, is in the civil engineering classes of work.

For UK jobs the standard method of classification is normally used because there are computer programs available to aid billing which are based on the A–Z classification. For overseas work a non-CESMM method of billing is used, which allows the classes of work adopted to follow the logical building order.

Number of items. Some civil engineering bills of quantities contain upwards of a thousand items because many different types of operations over many different structures are involved. Where possible, an effort should be made to keep the number of items to no more than they need be. This helps to reduce the work involved in measurement throughout the contract; but departures from the standard method may make the estimator's task more difficult and so should be kept to the minimum necessary.

The question of how detailed the billing should be depends on the nature and size of the works. What is to be measured for payment can vary widely. For instance, in a contract for the construction of a dam, some minor gauge house might be billed as a single lump sum item; the drawings and specification providing all details of what is required. Often where there are repetitive structures, such as access chambers to valves on a pipeline, these too can be billed complete by number.

In civil engineering it is quite common to bill items, such as standard doors simply by number, the specification describing what is required including the frame, priming and painting, and the type of door furniture required. If a special door is required, such as for the front entrance, again this is shown on the drawings and specified in detail; so the item in the bill appears as 'Front entrance door ... 1 No. ... '.

Where methods of measurement depart from the ICE standard method, this must be made clear in the bill. Although the standard method permits the description of an individual item to make clear it is not measured according to the standard method, it is better to group such items together. Either they can be grouped under some appropriate sub-heading, or it may be decided that certain types of work throughout the bills are not to be measured according to the standard method. When this policy is adopted, a statement must appear in the preamble to the bills of quantities (see below) saying such as 'Painting of metalwork is not measured separately and is to be included in the rate for supply and fixing of metalwork'. To prevent errors, a sub-heading before metalwork items should repeat this briefly, for example, 'Following items including painting'.

15.5 Accuracy of quantities: provisional quantities

In preparing the bills, the quantities should be accurately taken off drawings in accordance with the method of measurement. The quantities billed should not contain hidden reserves by 'over-measuring' them when preparing a bill. There may be a temptation to do this when, for instance, billing the trench excavation for a pipeline. But if the engineer increases the length at greater depth and decreases that at shallow depth to compensate, he may give the contractor a false impression of the nature of the work. It needs to be borne in mind that sometimes it is the practice to 'agree bill quantity' for an item for payment if there is no obviously large variation from what the drawings show. Hence quantities should represent a best estimate of what will occur, in order to be fair to both contractor and employer.

The problem of rock. A problem occurs when billing rock which may be suspected but whose incidence is not known – as in the case of a long pipeline where it is impracticable to sink enough borings in advance to discover the depth and extent of rock everywhere. Sometimes a provisional quantity is put in for rock, but if the extent of rock is not known, the problem is to decide what provisional quantity is to be put in the bill? Also how can the tenderer price such an item when the actual quantity to be encountered is only 'provisional'. Instead, it is suggested, a *provisional sum* should be included in the bill for rock excavation, and a price for excavating rock should be agreed with the contractor, if rock is encountered.

However, a consequent problem is that, if rock is encountered, it will almost certainly delay the work, so the contractor will put in a delay claim. Despite this, there is much to be said for negotiating a rate when rock is encountered, because widely different methods – and therefore costs per unit excavated – will apply according to the nature and direction of bedding of the rock encountered (see Section 15.7).

Provisional quantities for other matters should likewise only be used with care. They should relate to something known to be required, the quantity being a reasonable judgement as to what might be required. This could apply to such matters as bedding pipes on soft material, or bedding and haunching pipes in concrete, or fully surrounding pipes in concrete where the actual extent of such work depends on the site conditions encountered.

15.6 Billing of quantities for building work

Quite complicated buildings often form part of a civil engineering project, for example power station buildings, pumping stations, stores, administrative offices or laboratories. The CESMM gives units of measurement for some common building operations, but the nature of building work is so diverse that, in practice, many more items than shown in CESMM will be found necessary.

The CESMM will usually be found suitable for billing all work required to complete the framework, walling, cladding and roofing to buildings, and such matters as pipework, roads, sewerage, landscaping and fencing. Other building items primarily cover the interior finishes, carpentry and joinery, and other miscellaneous matters. These can be measured by some simple method a civil contractor will understand, since he will have experience of building work as well as civil engineering. It is not necessary to follow all the details of the standard method of measuring building work, for example the many 'extra overs' listed in that method for brickwork. The preamble to the bills of quantities should then clearly state which classes of work (i.e. 'trades') are not measured in accordance with the CESMM.

For matters which it is usual to let out to specialist sub-contractors, such as terrazzo floorings, balustrading, ceramic tiling, etc., either lump sums can be called for if the drawings and specification define everything required, or provisional sums can be inserted.

15.7 Some problems of billing

Excavation

Apart from excavation by dredging or for 'cuttings'; the CESMM distinguishes only between 'excavation for foundations' and 'general excavation' (listed in that order). However, the more logical order should be adopted of billing general excavation to a stated level (the 'final surface' for that item) followed by excavation for foundations below the 'final surface' for the general excavation. This can result in items referenced 'E400' in accordance with CESMM, preceding those referenced 'E300'; but these references should not be changed.

If the general excavation has to be taken down to two different levels, that is to a 'stepped' formation, then under the CESMM method it is billed as one item to the lower of the two 'final surface' levels. If an attempt is made to bill it as two excavation items 'banded horizontally', one below the other, sundry complications occur which are best avoided as they can cause confusion. Nor should it be taken as two separate items, the depth of each being measured from ground surface, because it is not excavated in this manner when the areas are adjacent.

Rock excavation has to be itemized separately from other materials, the volume of rock being measured independently. Usually neither the quantity nor the depths at which rock will be encountered will be accurately known; but the quantity should be estimated on the basis of the geophysical data available. The latter must be supplied to tenderers to permit them to make their own judgement as to the depth and extent of rock likely to be encountered.

The definition of 'rock' presents difficulties, but it must be stated in the preamble to the bills of quantities. Geophysical data may occasionally permit a given 'rock' to be defined, but in most cases rock is probably best defined according to the method of excavation. Unfortunately methods for removal can vary greatly, but for specification purposes three methods can be distinguished:

- use of explosives;
- use of hydraulic hammers or compressed air-operated tools;
- use of mechanical rippers (in open excavations).

It is usual to combine the first two methods to define rock by defining them as – 'Rock is material requiring to be loosened or broken up in situ by use of explosives or hydraulically operated rock hammers or compressed air-operated rock breaking equipment before being removed'. From the contractor's point of view this is not entirely satisfactory since it would exclude payment for rock he can get out with a suitably powerful digger able to cope with hard bands, albeit with difficulty and at a slow rate, involving substantial extra cost. Hence mechanical ripping forms a third category which may warrant separate measurement where hard bands of material are encountered that cannot be broken up by scrapers or the normal bucket excavator, but do not qualify as rock. Measurement of rock excavated for valuation is not easy; it is best done by a member of the resident engineer's staff and the contractor's staff viewing the excavation together in order to agree on the rock volume.

Working space

Contractors often claim payment for additional excavation to provide working space, despite the fact that most contracts and methods of measurement clearly state that only the volume vertically above the limits of foundations will be measured for payment. Therefore if some exterior tanking or rendering to a basement is required, it is advisable to repeat in the item for this that the contractor must allow in his rates for any working space he requires.

Pipelines

Trench excavation for pipelines is covered piecemeal in Classes I, K and L of the standard method (Class J covers provision of fittings and valves). Trench excavation to pipe invert level is included in the supply, laying and jointing of pipes per linear metre in Class I. Excavation below that for bedding is included in the supply and placing of bedding material, also per linear metre, in Class L. Extra excavation for manholes is included in the rates for manhole construction in Class K. Rock is an extra item payable per cubic metre in Class L. All pipework excavation items include backfilling.

Excavation of joint holes is not specifically mentioned so needs to be specified as included in the rates.

If the standard method of measurement is not used, it can prove simpler to take excavation (including backfilling) separately from pipe supply and laying. The maximum and average depth of trench, including any depth required for bedding, is stated for any given length of pipeline and is taken for payment per linear metre. Excavation for joint holes should be stated as not measured but included in the rate for trench excavation. The drawings should show the standard trench widths taken for payment, and the depth of any bedding. Rock is paid for as an extra over per cubic metre within the payment limits, the rate to include for overbreak and backfilling thereof. Bedding, haunching and surrounding are measured per linear metre for supply and placing.

Thrust blocks for pipelines have to be constructed against vertically cut undisturbed ground. It avoids argument if items for thrust blocks to dimensions shown on the drawings are followed by an extra-over item for trimming sides of excavation adjacent to thrust blocks to the vertical, including any backfilling between a thrust block and the vertical excavated face with concrete. Under the CESMM thrust blocks are measured per cubic metre inclusive of concrete, formwork, reinforcement, etc. For large pipes requiring major blocks it may be better to deviate from the standard method by treating these blocks as structures in their own right.

Earthwork construction

Earthwork construction is measured as the net volume as placed. The source of the filling should be stated. All information available about the nature of the proposed fill material should be supplied to tenderers so they can make their own estimate of the bulking factor of loose filling, its weight per unit volume loose and when compacted, etc. When the filling is to be obtained from a borrow pit, information concerning the extent and characteristics of the borrow pit material and its location should be provided in the tender documents.

If specified material for filling is to be obtained from *selected* material from a borrow pit, the removal or set aside of unsuitable material from the borrow pit has to be included in the rate for filling. It may also be necessary to include re-handling of the unsuitable material in order to put it back into the borrow pit. It is impracticable to measure the unsuitable material because some may be worked around and left in situ. Hence it is important to define what the rates for placing filling obtained from a borrow pit are to cover. Failure to include any necessary double-handling of unsuitable material can result in a large claim for extra payment from the contractor. It is advisable to give separate items for stripping overburden from the borrow pit, and an item for reinstatement of the borrow pit. The specification should set out all the requirements needed for reinstatement which it should then be possible to bill as a lump sum item for pricing. One point to note is that when 'suitable material' has to be taken from a borrow pit, it may be helpful to specify instances of 'unsuitable material' also.

Concrete

Concrete in situ is measured in the CESMM as two operations: (i) supply according to various quality grades; and (ii) the placing of concrete according to its location in beams, columns, slabs, etc. This suits the modern practice in the UK and similar developed countries where widespread use is made of ready-mix concrete delivered to site. The totals of concrete in the various grades must therefore sum the same as the relevant placing items per grade.

If the CESMM is not used, the supply and placing of concrete can be itemized together, the grade of concrete being stated. This reduces the number of items in the bill and simplifies measurement for valuation.

If holes in flat slabs have to be left open for some other contractor, such as a separate plant contractor, then the bill should include items for the supply and fixing of temporary covers to them to prevent accidents.

Brickwork

The standard method (CESMM) measures brickwork per m^2, the thickness being stated. It does not classify brickwork according to height above ground, nor separate out cavity walling in brickwork. It is therefore simpler to adopt non-CESMM billing, that is, measuring external brick walling from footings to d.p.c. level; and thereafter in one- or two-storey heights; separating cavity wall brickwork (including provision of wall ties) from solid brickwork; and either including provision of facing bricks and 'fair-face jointing' in the cavity walling, or allowing this as an 'extra over'. The specification must set out in detail what is required in respect of type of bricks and blocks to be used, and wall ties, surface finish and type of joint, etc. The bill item descriptions should also repeat what is to be included in the price quoted. Where ventilators to walls are required, these can be itemized inclusive of 'building in'. Masonry or precast concrete cladding needs to be billed separately, and angle supports, cramps and dowels (often of stainless steel) should be included or itemized separately. Other items separately billed will be d.p.cs, lintols, brick arches, etc.

15.8 Use of nominated sub-contractors

Some of the problems associated with the contractor's use of sub-contractors have been described in Sections 7.7 and 7.8. This section deals with additional problems that can arise with the use of nominated sub-contractors.

When a sub-contractor is nominated in a contract it is important that the specification sets out what services the contractor must supply to the nominated sub-contractor. These may include – providing access, off-loading materials, providing electrical power, scaffolding, cranes, etc. and permitting use by the sub-contractor's men of the contractor's canteen and welfare facilities. It will also be necessary to define how much notice the sub-contractor must be given before he is able to deliver equipment or start work, when he can undertake his work and how long it will take.

However, the actual terms of the sub-contract have to be decided between the contractor and sub-contractor, and there can be instances where they cannot agree. The sub-contractor may refuse to indemnify the contractor 'against all claims' and costs, etc. as required by Clause 3 of the Form of Sub-contract (described in Section 7.8) and the contractor may then refuse to place the sub-contract. It is true that, under Clause 59(1) the contractor has to have a 'reasonable objection' for refusing to employ a nominated sub-contractor, but this constraint is of little value in practice. There can also be refusal of either party to accept the other's terms for payment – although Clause 59(7) of the ICE Conditions endeavours to protect a nominated sub-contractor by providing for direct payment to him if the main contractor fails to pay him.

If the contractor or nominated sub-contractor will not enter a contract between them for any reason, the nomination fails. The engineer then has to nominate another sub-contractor, or ask the contractor to do the work himself or find his own sub-contractor. This mixing of responsibilities between the employer and main contractor for the performance of a nominated sub-contractor leads to frequent disputes. If the nominated sub-contractor fails to do the work or goes into liquidation, the employer must take action without delay, since the main contractor has no duty to carry out the work himself and will claim for any delay in getting it done. Further, if the sub-contractor's work proves unsuitable or not fit for its purpose, the main contractor will deny responsibility (unless he was expressly charged in the terms of the sub-contract to take such responsibility – which is unlikely) since he had no choice in the selection of the sub-contractor or the product.

The extensive disputes which have arisen over the years on these matters, more particularly in building work with its wider use of nominated sub-contractors, has led many engineers to take the view that use of nominated sub-contractors in civil engineering should be avoided wherever possible.

The problems most often arise from late completion or defective performance by a nominated sub-contractor which affects the contractor or his other sub-contractors. A possible way to avoid this is to specify the work in detail, providing for its payment by measure under appropriate bill items. Or, if specially skilled or experienced workers are required to do the work, there is no reason why this cannot be specified by calling for particular craft skills and requiring evidence of same. Alternatively, if only one firm can provide the special techniques required, the letting of a separate contract for such work can be considered. Although the organization of separate contracts has to be efficiently managed (as described in Section 5.6), the advantage is that the engineer retains direct control over the specialist's work and can act to avoid or solve problems arising from late delivery or defective work. Where a nominated sub-contractor is to supply only certain items, the employer may be persuaded to order the materials direct and store them in advance of being needed; or the contractor may be instructed to order them early and can be paid for offloading and storing them on site under items provided in the bill of quantities.

15.9 Prime cost items

When a nominated sub-contract is intended but not yet chosen, a prime cost item can be inserted in the bill of quantities by the engineer to cover the work and/or materials to be supplied by the nominated sub-contractor. The sum entered by the engineer is that which he estimates will cover the sub-contractor's charges; the actual charge made by the sub-contractor being refunded to the main contractor. Two additional items are added which tenderers can price. One consists of a lump sum to cover the *general facilities* the contractor is to provide for the sub-contractor; the other, expressed as a percentage of the prime cost, is to cover all the contractor's other charges and his profit.

The *general facilities* as defined in CESMM are deemed to cover – access; use of scaffolding, hoists, contractor's messrooms and sanitation; space for any office and storage the sub-contractor sets up; together with light and water. If more facilities than this are required, the CESMM requires this to be expressly stated. Such extra facilities are often required, such as provision of power, labouring assistance, use of crane, etc., and these have to be defined. If, however, the sub-contractor supplies materials only, the lump sum is deemed to include the contractor's unloading, storing and hoisting of materials delivered.

One practice to be avoided is to permit a sub-contractor chosen by the contractor to submit his quotation only to the contractor. This could lead to the sub-contractor including in his quotation for doing some work that is paid for elsewhere in the contractor's contract, resulting in double payment to the contractor. A preferable approach is for the engineer to call for quotations from sub-contractors to be submitted to him.

15.10 The preliminaries bill and method-related items

A preliminaries bill lists items which apply to the contract as a whole, such as insurance of works, offices for the resident engineer, provision of laboratory, surveying equipment, transport, telephone and tests on materials or the works. The units of measurement will be appropriate to the type of item, lump sum, or per week or month, or per number of tests, etc. Sometimes an item needs to be split into two parts, such as a lump sum for provision of the engineer's site office, with a second item for its maintenance per week or month. Such items listed by the engineer in the bill must be supported by descriptions in the specification stating exactly what the contractor is to provide. See Fig. 15.1 which shows part of the first page of a preliminaries bill drawn up according to CESMM.

Temporary works

The engineer may list temporary works the contractor has to provide, such as access roads, a temporary sewage treatment plant and similar. The listing of such temporary works permits the contractor to put a price to them, which may be to his advantage. Insurance is costly, so that if a tenderer prices this

CLASS A: GENERAL ITEMS

Item no.	Item description	Unit	Quantity	Rate £	Amount £
	CONTRACTUAL REQUIREMENTS				
A110	Performance bond	Sum		—	16 000.00
A120	Insurance of works	Sum	)		
A130	Insurance of constructional plant	Sum	)	—	33 500.00
A140	Insurance against damage to persons and property	Sum	))		
	SPECIFIED REQUIREMENTS				
	Accommodation for engineer's staff as specification Cl.112				
A211.1	Provide and erect	Sum		—	9 500.00
A211.2	Maintain	Wk	104	200.00	20 800.00
A211.3	Remove	Sum		—	1 000.00
	Services for engineer's staff as specification Cl.113-116				
A221.1	Provide transport vehicles	Sum		—	17 500.00
A221.2	Maintain and service transport vehicles	Wk	104	70.00	7 280.00
A222	Install telephone and fax	Sum		—	1 000.00
	Equipment for use by engineer's staff as specification Cl.117-118				
A231.1	Provide office equipment	Sum		—	10 000.00
A231.2	Maintain office equipment	Wk	104	30.00	3 120.00
A233.1	Provide survey equipment	Sum		—	12 000.00
A233.2	Maintain survey equipment	Wk	104	10.00	1 040.00
A239	Insure telephone, fax, office and survey equipment	Sum		—	Included
	Attendance upon engineer's staff ... ... etc				

Notes: The above assumes water supply, sanitation and electrical supply are specified under accommodation; they could have been separately itemized. Fuel for vehicles, and telephone charges would be paid by engineer. Removal of telephone, office and survey equipment could have been itemized but is ignored.

Fig. 15.1. Typical example of the first page of a preliminaries bill, drawn up according to CESMM Class A requirements, and priced by tenderer

item, he can be reimbursed his expenditure on it as soon as he shows evidence of obtaining it. He does not have to wait for its reimbursement as he would have to if he spread the cost over the constructional items. Also, a priced item for such as the site sewage treatment works to be provided by the contractor is of advantage to the employer, since payment for such works can be withheld if the plant is not as specified or does not work properly. However, the pricing of a tender is in the tenderer's hands and he does not have to put a price to any of the items listed by the engineer in the preliminaries bill. He can mark them 'included' meaning the cost of meeting the item requirements is included in his bill rates for the construction items, or he may enter a low figure.

Items added

A tenderer may sometimes add an item which is not in the list set out in the preliminaries bill. For instance, he might wish to separately price some especially expensive temporary works equipment, such as steel shuttering. However, the employer may have laid down in the Instructions to Tenderers that 'no items shall be added to the bill of quantities'. The employer can refuse to consider such a tender; but if it is the lowest tender received the employer may decide nevertheless to consider it. This depends on the rules under which the employer himself operates, such as the standing rules of a public authority, or government regulations. Normally, however, an extra item or two added by a tenderer in the preliminaries bill would not be taken as invalidating a tender. Some contracts specifically allow this by writing in the preliminaries bill 'Other items added by contractor…'. If a tenderer does add such items they should be discussed at tender negotiation stage to agree how they should be paid. For example, payment of a lump sum for steel shuttering might need to be agreed as a certain percentage on delivery, the balance on completion of its use.

Method-related items

The ICE standard method of measurement of 1985 recognized that items of the foregoing kind could be added by tenderers. It called them 'method-related' items, though they are not confined to construction methods but include organizational measures as well. The CESMM, 3rd edition, lists over forty such matters that a tenderer can add, covering such as – accommodation (offices, stores, canteen, etc.); services (water, power, site transport, welfare, etc.); plant and temporary works of many kinds, 'supervision and labour', and also permits a tenderer to add other method-related items not in those listed. All such method-related items have to be priced as lump sums, defined as either **fixed** or **time related**. If fixed, the lump sum is only payable when the work itemized is completed. If 'time-related' the payments are spread out over the time taken to achieve completion of the work covered by the item (see clarification in Section 16.4). Clearly some items, such as supervision, site transport, welfare, should not be designated as 'fixed' as there is no definable time when they could be said to be completed, other than the end of the contract. A tenderer has to define exactly what any item added by him covers, and whether it is fixed or time-related. Figure 15.2 shows some typical method-related items entered by a tenderer.

CLASS A: GENERAL ITEMS

Item-related	Item description	Fixed charges £	Time-related charges £
	METHOD-RELATED CHARGES		
A311.1	Offices	39 000.00	—
A311.2	Maintain offices	—	3 900.00
A314	Stores	9 000.00	—
A315.1	Canteen	11 000.00	—
A315.2	Maintain canteen	—	5 200.00
A320.1	Services, install	15 000.00	—
A320.2	Services, maintain	—	10 400.00
A320.3	Services, remove	1 500.00	—
A330.1	Plant, provide	20 000.00	—
A330.2	Plant – maintain 60 weeks	—	30 000.00
A330.3	Plant, remove	2 000.00	—
A333.1	Piling plant, provide	14 000.00	—
A333.2	Piling plant, remove	1 000.00	—
A341.1	Drilling plant, provide	3 500.00	—
A341.2	Drilling plant, remove	1 000.00	—
A356.1	Provide pumping plant	2 000.00	—
A356.2	Remove pumping plant	500.00	—
A357	De-watering	—	10 000.00
A363.1	Provide piling	25 000.00	—
A363.2	Remove piling	8 000.00	—
A370	Supervision and administration	—	104 000.00
	Total fixed charges	152 500.00	
	Total time-related charges		163 500.00
	Total to summary		316 000.00

Note: Some of the fixed charge items need more precise definition so that it can be known when they are completed e.g. services, plant etc. The time-related item A330.2 cannnot be paid over 60 weeks at £500 per week, it must be paid over the contract period.

Fig. 15.2. Typical method-related items entered by a tenderer – one item wrong

Division of items in the preliminaries bill

The standard method has five main divisions or categories of items that can be put in the Class A preliminaries bill:

1. 'contractual requirements' (bond and insurances);
2. 'specified requirements';
3. the 'method-related charges' referred to above which the tenderer is to insert;
4. 'provisional sums';

5 and 6. 'Nominated sub-contracts', which include work done on site, and work, such as manufacture, done off site respectively.

The 'specified requirements' (2) cover accommodation and services for the engineer's site staff, tests on materials, etc., and a range of temporary works that the engineer might wish to itemize.

The difference between temporary works the engineer itemizes as 'specified requirements' under Division 2 of the Class A bill, and the temporary works which a tenderer adds as 'method-related' items under Division 3 should be noted. The former have to be fully specified by the engineer in the contract; the latter do not, being left to the tenderer to describe. Thus if the contractor is required in the specification general clauses to construct some temporary access road, then if the engineer itemizes it as a 'specified requirement' in Division 2 the details of it must be fully described in the specification or contract drawings. If the engineer does not know how the access road should be constructed because he does not know what traffic the contractor will put on it, then he should not itemize it in Division 2 but leave it to the contractor to add in Division 3 as a method-related item, if he so wishes.

It is important to follow the standard method requirements exactly, or problems of interpretation leading to claims from the contractor may arise. Of course the contract can expressly state that items in the Class A Preliminaries Bill are not drawn up in accordance with the standard method; but then care has to be taken to define what each item entered covers so there is no ambiguity.

Problems with Civil Engineering Standard Method of Measurement

The whole concept of payment for temporary works as set out in CESMM can be called into question, as it creates potential ambiguities. The engineer may choose not to itemize any temporary works under 'specified requirements' because he leaves such works for the contractor to decide. But the contractor may maintain that the list of temporary works given in CESMM A.2.7 (such as traffic diversion, access roads and de-watering) entitles him to payment for those works on the same principle as – when an item which CESMM lists for measurement is found missing – the item has to be added to a bill (see end of Section 17.2). To avoid this ambiguity the preamble to the bill should state that Class A items shall be measured only to the extent they are included in the contract at the time of the award; thus fixing the temporary work items measured.

Another difficulty arises with method-related items. CESMM clauses state that a method-related charge does not bind the contractor to use the method defined (Clause 7.5); is not subject to admeasurement (Clause 7.6); and is not to be increased or decreased for any change of method adopted by the contractor (Clause 7.8). But when the engineer orders a variation of some permanent work, the contractor may claim that bill rates for similar work do not apply, because the temporary works associated with that work have changed but the method-related item of charge remains fixed. This can raise debatable issues concerning method-related charges which are defined as not subject to admeasurement and they need bear no relationship to actual methods the contractor uses.

Under ICE Conditions (Clause 14(7)) the engineer is only required to state why a proposed method by the contractor fails to meet the contract requirements or would be detrimental to the permanent works. It is left to the contractor to decide what method he will adopt to gain the engineer's consent. Hence, if the engineer has no reason to specify a particular method, he should avoid mentioning any lest this be interpreted as a 'specified requirement' as discussed above. Also acceptance of a method-related item in a contract does not imply the engineer has given his consent to the method stated. The preamble to the bill may need to make this clear.

15.11 Adjustment item to the total price

An **adjustment item** is an addition or deduction a tenderer makes to the final total of his prices entered in the bills of quantities. The CESMM permits an adjustment item as a lump sum addition or deduction, paid by instalments in the same proportion as the total payments to date, less retention, bears to the total of billed prices (see Section 16.4). The addition or deduction is not to be exceeded, and the full amount is to be allowed when a certificate of substantial completion for the whole works is issued.

In contracts which do not follow the CESMM a tenderer may be free to add an adjustment item to his tender – or in fact add any additional item for which he submits a separate price. His tender is only an 'offer' so he is free to offer his price in any way he likes. The employer can, of course, lay down rules that he will not entertain any offer that is not priced as he instructs, but this is a rule for himself. The tenderer has to run the risk that his non-conforming tender will not be considered: but this is rather unlikely to happen if his bid is the lowest. Thus, instead of inserting a lump sum addition or deduction as required by the standard method, he can insert an adjustment item which comprises a percentage reduction (or, more rarely, addition) to be applied to all his billed prices. Sometimes this practice is actually invited by the employer who invites tenders for two separate contracts simultaneously, and provides a special item in one contract for the contractor to quote his reduction of price (if any) if he were awarded *both* contracts.

An adjustment item as such is usually added by a tenderer when – after having had all the items in the bill priced and totalled – he looks at the final total so derived and decides to increase or decrease it. This is his commercial decision. He will have made a check estimate of the cost of the whole contract in an entirely different manner from that obtained by totalling the priced quantities in the bills. This can be done, for instance, by costing the total materials and estimated labour and plant to be used on the job, and adding a percentage for overheads and profit. In the light of his findings and taking into account other factors, such as risk, need for more work or the likely competition from other tenderers, etc., the contractor may decide to add or subtract an adjustment figure to the total of billed prices. He could, of course, select certain bill items whose rate or price he could alter to make the adjustment, but this could be risky if more or less work under such items should prove necessary.

15.12 Preamble to bill of quantities

There must be a preamble to the bill of quantities in which is stated, among other things, the following:

- the method of measurement used in preparing the bills of quantities;
- if the CESMM is used, the edition which applies and which parts of sections 1–7 (which cover general instructions, etc.) are to apply;
- the classes or types of work which are not measured in accordance with the CESMM;
- provisions with respect to 'method-related' items;
- provisions with respect to any 'adjustment item' to the total of billed prices;
- payments to be made in respect of prime cost items;
- the definition of 'rock';
- if CESMM is used, identification of bodies of water on or bounding the site;
- if no price is entered against an item that it will be assumed that no payment is to be made under that item.

The provisions with respect to the method-related items inserted by the tenderer may need further amplification added before the contract is awarded, to clarify such matters as the method of payment of such items.

15.13 List of principal quantities

The CESMM requires that 'the principal components of the works with their approximate estimated quantities shall be given solely to assist tenderers in making a rapid assessment of the general scale and character of the proposed works prior to the examination of the remainder of the bill of quantities'. This list is to precede the preamble to the bill of quantities. It is difficult to understand how this requirement could be of any real value to a serious tenderer. It would in any case be a subjective selection by the engineer of 'the principal components'. It is the tenderer's responsibility to select those components that are most significant to him in terms of cost, quantity or difficulty. The early part of the specification should describe the nature, magnitude and output or size, etc. of the principal components of the works so that the extent of the works required is defined and readily appreciated. This is also useful if the cost of the contract is to be of value for cost analysis purposes in the future.

16

Interim monthly payments

16.1 Handling interim payments

Under the ICE conditions, regular payments based on the quantity of work done during the previous month, must be made by the employer to the contractor at monthly intervals. The amount of work done is measured by the engineer under the contract, and valued in accordance with the terms of the contract. The engineer then issues a **certificate of payment** showing the amount which the employer must pay to the contractor. Occasionally other intervals for payment may be agreed to suit accounting periods, for example, payments at 4 weekly intervals. Sometimes it is agreed that two out of every 3 monthly payments are approximate valuations of work done; thus only the quarterly payments are based on a detailed measurement of work done.

While the ICE conditions and other standard forms have long had such payment terms, many bespoke forms, such as those for sub-contracts, have different terms for payment. The *Housing Grants Construction and Regeneration Act 1996* (see Section 1.6) now requires that all UK construction contracts contain terms allowing regular payment and means of assessing the amount due. The Act also outlawed pay-when-paid clauses.

The ICE conditions require certification by the engineer within 25 days of the contractor submitting his account with payment made within 28 days. If payment is late the contractor can charge interest on the overdue payment at 2 per cent per annum above bank rate for each day late.

These are onerous requirements. During the period of 28 days, the resident engineer has to check the contractor's account, amend it as necessary and forward it to the engineer whose contracts department may need to check it. The engineer then issues his certificate and sends it to the employer. If the employer is a government, local government or other statutory authority it may need more than one person to authorize payment, and the account then has to be passed to the paying department of the authority. The stipulation of

28 days represents 20 working days which may seem unreasonably short. A longer period may sometimes be appropriate. Few, if any, contractors pay their suppliers' accounts in 28 days.

Under FIDIC conditions for international work the engineer has 28 days within which to issue a certificate for interim payment, and the employer a further 28 days within which to make payment.

As a consequence of the short time period for payment under ICE conditions, the resident engineer must try to agree quantities, or the value of work done, with the contractor before he draws up his account. The contractor will need to be warned that if he submits quantities or items for payment which have not been prior agreed, there will be no time for the resident engineer to hold discussions on them; he will substitute his own measurement or amount payable in lieu.

The contractor should be required to submit at least two copies of his interim account in a standard form which is set out in the specification to the contract (see Section 13.7). The account should be in a form agreed beforehand, having three quantities columns showing 'Paid last certificate', 'Addition this certificate' and 'Total to date'. The account will no doubt comprise computer printouts which need checking to ensure bill rates are as tendered and the charges are arithmetically correct. Where the resident engineer does not agree with items charged he should mark both copies before sending one copy to the engineer. Subsequently the engineer may notify the resident engineer of any further amendments made by him (or his contracts department), which must be entered on the resident engineer's copy, and he must send a letter to the contractor notifying the detail of all amendments made to his account.

An important point the resident engineer should remember is to mark on all accounts the date of receipt.

It should be remembered that monthly payments are only interim in nature and can be adjusted for in a subsequent month if found to be wrong. Only the final account decides the amount due under the contract.

16.2 Agreeing quantities for payment

The way in which the resident engineer should measure quantities has already been described in Section 13.6. In that section the importance was emphasized of making clear what has been measured, and what has been agreed for payment – with sketches as necessary.

It is strongly advisable that the resident engineer should take the lead in assessing final quantities. He or a member of his staff should supply quantity calculations to the contractor (or his quantity surveyors) and request agreement. If the opposite method is adopted of the contractor supplying his quantity calculations for the resident engineer to check, the resident engineer may find it difficult or impossible to find out why some contractors' quantities differ from his own. This is especially so if the contractor's quantity calculations are done according to some computer package which is unfamiliar to the resident engineer, or if no sketches are provided by the contractor to explain how his quantity has been calculated.

Some items will need measurement in the field, such as trench depths for pipelines, excavation and mass concrete to foundations, etc. This may be done by the resident engineer's inspector who agrees the measurement with the appropriate foreman, a written advice note being sent immediately to the contractor giving the agreed figures. Sometimes joint measurement is arranged. Other quantities can be taken from drawings if no variation has occurred. Where an item of work is only partly done, a rough estimate of the proportion done should be agreed with the contractor. Sometimes all but a minor aspect of an item is completed, for example, final painting of valves in a valve chamber. It is simpler to certify the item in full rather than making some trivial reduction, making a note to remind the contractor if he fails to complete the painting.

If work has been done so badly that it cannot be accepted, no payment should be certified for it. If the contractor has agreed to do some remedial work that will make it satisfactory, some partial payment can be made. All depends on the circumstances. With a reputable contractor and a responsible agent there is no reason to assume that verbal promises will not be carried out. With good contractors most agreements are verbal anyway, and any paperwork is only for the purposes of record.

Some agents leave the contractor's quantity surveyors to prepare the contractor's accounts. When the quantity surveyors are approached and asked not to repeat in every account items of claim previously struck out as being invalid, they may reply they have no authority to remove an item once it has been put into an account. Continued practice of this kind means that, as the measurement gets larger, the resident engineer may have to make an increasing number of corrections to items, page totals, and bill summaries. If the engineer is unable to get the practice altered – if necessary by direct approach to the contractor's senior personnel – it at least eases the problem to insist, from the beginning, that all extras, added items and claims are billed on a separate sheet, or added at the end of each bill.

16.3 Payment for extra work, dayworks and claims

When extra work has been ordered then, if any of such extra work has been done by the contractor, there should be some payment for it in the next interim payment certificate even if the rates for such extra work have not been agreed. The reason for this is that the ICE conditions provide that work done shall be measured monthly and paid for within 28 days of the contractor submitting his account for same. Any dispute about the exact rate or amount for some extra should not therefore delay payment of what the engineer considers a reasonable amount. Otherwise the contractor could claim he was not paid the undisputed portion of his application within 28 days of submitting his account, and so might be entitled to claim interest on it for late payment (see Section 17.13).

Dayworks charges (see Section 13.8) which have been checked and agreed by the resident engineer will also need to be included in the next interim certificate for payment, in so far as the contractor lists them in his account. The contractor's account may, of course, list other dayworks charge sheets that have not been previously submitted to the resident engineer for checking. If there is time to check them, the resident engineer should include them for payment. Alternatively he might include a round sum 'On account of unchecked daywork sheets submitted', the sum being what he considers will at least be payable under them. However, he does not have to include any payment against a daywork account which provides insufficient information for him to check it.

Similar problems of the need to make partial or 'on account' payments can arise in respect of claims submitted by the contractor which may be valid in principle but not sufficiently detailed to support the full claimed figure. Nothing need be certified in respect of a claim insufficiently detailed for it to receive any consideration (ICE conditions Clause 53) but amounts should be included for payment where sufficient details are provided to justify some payment. All such partial or 'on account' payments proposed by the resident engineer should be drawn to the attention of the engineer when the contractor's account is forwarded to him.

Where the resident engineer has agreed rates for extras with the contractor he should draft an appropriate variation order which he sends to the engineer for checking and issue.

16.4 Payment of lump sums, method related items and any adjustment item

Unless there is some stipulation that a lump sum is to be paid in stages it only becomes payable when the whole of the work itemized under the lump sum is completed. However, fairness has to be applied when an item combines two operations and nothing is stipulated about staged payment. Occasionally a lump sum item reads: 'Provide and set up engineer's offices as specified and remove on completion' so in theory the item is not payable until the end of the contract. At the tender negotiation stage, agreement should be reached as to how the sum is to be paid; but if the matter has been missed then it is up to the resident engineer to suggest how the item should be paid – perhaps 80 per cent on set-up; 20 per cent on removal, or some other proportion. Lump sums for such as insurance can be certified for payment as soon as the contractor produces evidence of insurance.

Payment of method-related items (see Section 15.10) will depend on whether any special conditions have been laid down about them for staged payment, or if not, payment will be dependent on whether they are fixed or time-related. Fixed sums would be paid as mentioned above for ordinary lump sums, but problems can arise if the item description is imprecise. Thus if the contractor has added some method-related item for 'scaffolding' or 'site transport' as a fixed sum, it has to be further defined to relate to some specific scaffolding or specific transport; otherwise it is too vague to identify and cannot be paid until substantial completion.

When method-related sums are time-related they are paid in monthly instalments, but the proportion paid will depend on when completion of the relevant task will be reached. Thus if an item is completed when substantial completion is achieved, and the programmed time for this is 18 months then, 1/18th payment is added at the end of each month. But if it then appears that substantial completion will not be reached until month 20, the total payment due at month 7 is 7/20ths. This is irrespective of whether the delay in completion is due to the contractor's tardiness or to an authorized extension of the contract period. The reason for this approach is that method-related items are defined as covering 'costs not to be considered as proportional to the quantities of the other items'. If an extension of the contract period has been granted, then any claim for extra payment on that account is a separate matter to be decided by the terms of the contract. The lump sum payable under a method-related item remains unaltered.

An adjustment item in the form of a lump sum (see Section 15.11) is paid in the same proportion as the total payment due under other items less retention, bears to the total contract sum, less the adjustment item. If a percentage adjustment has been quoted, this is applied to the total amount payable under bill items and variations. In both cases the retention money is deducted after adding in the adjustment item.

16.5 Payment for materials on site

The ICE conditions and similar permit payment to be made to cover part of the cost to the contractor of materials delivered to site but not yet built into the works. This can ease the contractor's cash flow situation and is of advantage to the employer in encouraging early supply of materials so that unexpected shortages or late deliveries are less likely to hold up progress. In contracts that contain such a provision, tenderers can be expected to reduce their prices in anticipation of the expected financial benefit.

Certifying payment for materials on site is left to the discretion of the engineer. Under ICE conditions Clause 60(2)(b) he has to certify such amounts (if any) as he may consider proper, not exceeding a percentage of the value as stated in the contract. In this he may need to act carefully because, even though material has been delivered to site, it might still remain the property of the supplier until he has been paid for it by the contractor. If the supplier falls into dispute with the contractor, or goes into liquidation, the materials he supplied might be reclaimed by him or his receiver. Before certifying any payment for materials the engineer will need to be reasonably certain that the contractor does own them.

In deciding what should be certified for materials on site, the resident engineer needs to check they comply with the specification, are properly stored or protected, and will not deteriorate before use. The amount certified will depend on the nature of the material and also the circumstances of the contractor. If the contractor appears to be running into financial difficulties or shows signs of being unable to complete the contract, what should be certified for materials on site needs careful consideration by the engineer. The prospective value to the employer of the materials paid for, needs then to be assessed in the light of the situation, allowance being made for any deterioration that might occur if there is a delay in their incorporation into the works. Reinforcement or structural steel left out too long in the open may rust to the point of scaling; improperly secured items may get stolen; pipes left too long on verges to roads may sustain damage to their protective coatings; valves can be damaged by frost and so on.

16.6 Payment for materials manufactured off site

The ICE conditions also permit payment on account to be made for items which are manufactured off site (Clauses 54 and 60(1)(c)). This provision is intended primarily to cover mechanical or electrical equipment or prefabricated steelwork which the contractor has to supply for incorporation in the works. He will usually use a specialist manufacturer to supply such items. It is advantageous to the progress of the job for all such items to be manufactured and made ready for delivery in advance of the date planned for their incorporation in the works, hence payment for items manufactured off site encourages this.

However, only items listed in an appendix to the tender documents are to rank for on-account payment, that is, the contract pre-determines the equipment or plant to which the provision relates. Also two further conditions have to be complied with: (a) the equipment or plant must be ready for dispatch; (b) the ownership of it must be transferred from manufacturer to contractor, and then from contractor to employer. Clause 54 of the ICE conditions sets out the details of the procedure required.

Clearly before any payment on account can be made, the engineer or resident engineer will need to arrange for the manufacturer to be visited so that the plant to be supplied can be inspected to ensure it conforms satisfactorily to specification and all necessary tests before delivery. Evidence of the proper transfer of ownership, and sundry arrangements for storage, insurance, etc. will also be required.

16.7 Payment for manufactured items shipped overseas

When manufactured items have to be delivered to projects overseas, arrangements for staged payments will normally be provided for in the contract. Items will need to be inspected and tested at the place of manufacture, their loading to ships inspected, and inspected again when offloaded at the place of destination. If the civil engineering contractor is responsible for the supply of the items he must arrange for the loading and offloading inspections; if items are supplied under a separate contract the engineer will have to arrange the inspections. In either case, however, the engineer will need to ensure that such inspections are efficient, not only for the purposes of payment, but to ensure safe delivery because it may take weeks or months to replace an item lost or damaged.

Manufacturers normally only quote supply of equipment 'to dockside' or 'f.o.b.' (free on board), after which the carrier takes responsibility until he offloads. If equipment is not inspected at every stage, it may be impossible to know who is responsible for any damage or loss; leaving the employer to bear the cost of any replacement. The whole operation needs to be well organized if trouble is to be avoided.

16.8 Price adjustment

Some contracts contain a price variation clause in order to protect the contractor against the risk of rising prices due to inflation. Nowadays it is not usual for contracts in the UK lasting less than 2 years to contain such a clause. To calculate the amount due, the contractor either has to produce evidence of how prices have altered since he submitted his tender; or a formula which uses published indices of price changes is applied to the payments due to him under billed rates.

For contracts in the UK the price indices published by the Department of the Environment, Transport and the Regions, formerly referred to as the **Baxter indices** and which are relevant to the civil engineering industry, can be used according to a formula. The formula applies the indices via various weightings given to labour, plant, and specific materials – in rough proportion to their use in the works being built. Standard types of formulae are included in the ICE and other forms of contract. At each interim payment the formula is applied using the latest published indices to give a multiplier representing the change in construction prices since the date of tender. This is applied to the value of the work done and certified for payment during the month. The cumulative total of these monthly additions represents the allowance for price inflation for work done to date.

Most price variation clauses provide that the price adjustment ceases for those parts of the works for which a substantial completion certificate is issued, or for which contractual dates for completion are reached (including any extension given) – whichever is earlier. This provides an incentive for the contractor to achieve target dates.

Corrections may have to be applied if, as is often the practice, 'interim' indices are first published, followed later by 'final' values.

If no authentic indices are available for calculating price variation, as may occur on overseas projects, then price increases have to be directly calculated. Tenderers are required to list the basic rates of wages and prices of materials on which their tender is based. Checking the authenticity of these is usually done before signing the contract. Prices of materials may have to be checked by contacting suppliers direct, asking them to confirm what their price was at the date of tender. Any wage increases charged should have some authenticity, for example, be in line with inflation of cost of living or relate to some government or state policy for equivalent labour. Wage sheets and invoices for materials have to be supplied by the contractor as work is done: these are analysed to calculate the extra costs paid by the contractor. Sundry checks have to be applied of an auditing nature, for example, that the wages shown on the pay sheets were actually paid; that suppliers were paid what their invoices said; that the quantity of materials invoiced were used on the job and not on some other job; that invoices are not submitted twice over, etc.

In the hands of a competent and reputable contractor the checking work may be straightforward though very time consuming. Usually the engineer will draw up a construction contract which stipulates that 'only those materials named and priced by the tenderer will rank for price variation', in order to limit the number of items that have to be checked. The resident engineer will need to graph out the total price increases certified against total payments for work constructed, to ensure the increases follow a consistent pattern and are believably in line with known current price trends. The work is so time consuming and open to mistake or even falsification, that every effort is usually made to adopt some simpler and more reliable measure by means of a formula even if only a limited selection of indices is available.

16.9 Cost reimbursement

In recent years a number of employers have taken to using cost reimbursement terms for payment; often with a target cost (see Section 3.1(e)). In such contracts the contractor records his costs on an 'open book' basis so the employer can check and audit the books to confirm the validity of the costs to be paid to the contractor. Costs are normally recorded by computer and can be onerous to check as compared with a bill-of-quantities contract or other method of valuing work. Checking is thus often on a sampling basis, checking different categories of expenditure each month and concentrating on major costs with the intent of covering all important matters before a final account is agreed.

Some items of cost may be pre-agreed in the contract, or agreed subsequently as applicable to the whole contract, such as any fee, staff salaries, insurance, head office costs and use of contractor's own plant. The coverage of any such items must be clear to avoid possible duplication. Other costs will need to be checked in detail to ensure they were expended for the works and were reasonable in extent. Labour must be checked against wage sheets and the labour records held by the engineer's site staff. Hired plant can be checked against invoices, but the charge for contractor-owned plant, if not agreed in advance, may need checking by a specialist who has to allow for depreciation, maintenance and other costs and avoid any duplication of profit. Materials invoices must relate to actual materials used and discounts must be allowed for. Sub-contractors' quotations may need to be agreed in advance of the award of contract, to satisfy the employer that the prices are competitive and the terms acceptable.

Where a target cost has been set, it is necessary to keep a rolling check of costs incurred against the proportion of the target work done, so as to identify any significant differences and thus allow steps to be taken to investigate cost increases and look for means of reducing any over-run.

16.10 Retention and other matters

The retention money as stated in the contract (usually 5 per cent in the UK but subject to some maximum value), must be deducted from the total amount calculated as due to the contractor in interim certificates for work done. When a substantial completion certificate is issued, the retention held is halved for that portion or whole of the works to which the certificate applies, the amount so released being paid to the contractor. During the **defects correction period** (often termed the maintenance period) which is stipulated in the contract, the contractor undertakes to correct all matters listed by the engineer as needing remedial action. At the end of this period the remainder of the retention is to be released although a portion may be held back sufficient to cover any outstanding defects.

The performance bond which may have been provided by the contractor is normally not released until all defects have been dealt with, so the employer has some protection against default of a contractor in this respect. It is sometimes accepted that, after completion, any retention can be released by substitution of a retention bond.

The contractor's insurances will normally lapse once work, including remedial work, has ended. It should be remembered, however, that after substantial completion the normal contractor's insurance cover does not cover the works themselves since these have become the employer's responsibility.

The resident engineer must forewarn the engineer when substantial completion of part or all of the works is likely. If this is later than the contract period, or any extended period, liquidated damages may apply, as set out in the contract. If any are applicable, the employer may wish to deduct them from any payment due to the contractor, otherwise they may be more difficult to recover later. At completion the engineer will need to re-assess any extension of time due to the contractor and advise the employer regarding any damages due. Deduction of damages is not a matter for the engineer to certify, but for the employer to decide and apply.

17

Variations and claims

17.1 Who deals with variations and claims

The two parties immediately concerned with the issue of variation orders and the handling of contractor's claims are the **engineer** under the contract, and the **resident engineer**. While on overseas sites the resident engineer may have powers delegated to him to agree payments and to value variations (see Sec-tion 9.2) this would rarely be the case in the UK and, under ICE conditions, he cannot be delegated powers to settle the contractor's claims for delay or the cost of meeting unforeseen conditions in accordance with Clause 12(6). Also only the engineer has authority to issue the final certificate for payment, which of course, can include revision of any payments previously certified in the interim payment certificates.

In practice the resident engineer conducts the 'first stage' negotiation work on payment matters, digging out and recording all the relevant information, examining and checking the contractor's claims or justification of new rates he wants for varied work, and endeavouring to reach agreement with the contractor on what a fair rate should be under the terms of the contract. He reports all this to the engineer. If the resident engineer can get the contractor's agreement to a payment, which the engineer approves, this has the advantage that the contractor can have confidence that agreements he reaches with the resident engineer will not later be overturned.

The **employer** should normally be kept regularly informed of the progress and state of the contract. This information will include: major variations that the engineer has had to make; whether many claims have been put in by the contractor and what substance there is to them; and what the likely effect on the total cost of the contract will be. The employer may require to be consulted on any claim so that he can give his views on it before the engineer comes to a decision. Some contracts, such as the FIDIC 4th edition, specifically require the engineer to consult with the employer as well as the contractor before reaching a decision on a claim. But irrespective of whether such requirements exist, the engineer should always report major claims to the employer and allow both parties to put their views to him.

On very large projects where an ordered variation may incur heavy extra expenditure, it is advisable that the employer is involved in the issue of any order which incurs significant extra cost and many contracts stipulate this. On many such projects, including the Mangla project mentioned in Section 5.6, the engineer will report any proposed major variation to the employer for his agreement, with a technical report in justification. Variation orders can then be issued in two parts: Part I to issue the necessary instructions to the contractor; Part II to set out the terms of payment once discussed with the contractor. Thus urgent variations can be sanctioned by the employer and work can proceed. The employer needs a sufficiently large technical staff available to appraise the technical issues involved without delay. The advantage to the engineer is that the technical issues are thoroughly examined and solutions accepted before commitment to the very large sums which sometimes have to be sanctioned, and the advantage to the employer is that he is fully aware of the changes needed and their effect on the final cost.

While variations are normally decided and instructed by the engineer it is not uncommon for contractors to put forward ideas either to save cost or time. Indeed this is encouraged by value engineering procedures and can be a useful way of controlling or reducing the final price. The ICE conditions recognize the potential of such proposals and allow for sharing of any changes in value or time between both parties. Any proposals accepted must be instructed by the engineer before coming into effect.

Although the rest of this chapter mostly refers to the powers of the engineer under the contract, this must be taken as implying that the resident engineer must act similarly.

17.2 Payment for increased quantities

Re-measurement types of contract, such as those covered by the ICE 7th edition (Measurement Version), are let on the basis that the actual amount of work done is not expected to be exactly the same as that estimated from the contract drawings. The intention of the contract is that where a change of quantity requires no different method of working by the contractor and does not delay or disrupt his work then the billed rates still apply. However, Clause 56(2) of the ICE conditions recognizes that, if there is a considerable difference between the measured and billed quantity, the contract allows a review of the rate to ensure that a proper price is paid. If the engineer is of the opinion that a quantity has changed so much that 'any rates or prices (are) rendered unreasonable or inapplicable in consequence', then the engineer, after consultation with the contractor, can increase or decrease such rates or prices. The change in quantity has to be significant to justify an altered rate.

In some international contracts, any review of rates or prices is restricted to changes in individual quantities exceeding a given percentage. If it were intended that all changes in quantity justified a different rate, then the rates would largely become irrelevant. In practice therefore rates are seldom altered for what might be termed 'natural' variation of quantities. Most variations that are large enough to require re-consideration of rates stem from an instruction issued by the engineer or resident engineer and this is a different matter dealt with in Section 17.4. Occasionally an item gets missed from a bill, which the contract or method of measurement provides should have been measured. Rates for these must also be set by the engineer in the same manner, but without there being any instructed change.

17.3 Ordered variations

Many types of contract allow the engineer to order variations but, as mentioned in Section 8.4 his powers to do so are usually restricted to changes that are necessary or desirable for completion of the works, including any changes requested by the employer. By this means the employer, through the engineer, can obtain the result he wishes if his ideas and desires have altered since he awarded the contract. But if the employer wishes to introduce an entirely new piece of works or otherwise alter the basis of the contract, he can do this only by agreement with the contractor. The intention is that the works as contracted will still be built, that is, the same concept or result will be achieved but the detail may alter. This is essential, since the contract gives the contractor the duty and the right to carry out the contract works, and the contract will maintain those rights even if the works are varied.

The resident engineer may sometimes receive a request from an employee of the employer to make some addition. He should refer the request to the engineer who will need to consider whether the employer would agree to his employee's request, and whether it is within the engineer's power to instruct the contractor to make such addition. Obviously in matters of choice or no great cost the resident engineer will assent to reasonable requests, such as colour schemes for finishes. But sometimes during the finishing stages of a construction, the request may be for something expensive or which could delay completion, and it is then necessary to be sure the employer agrees with the request of his employee.

When ordering variations the ICE conditions set out the procedure to be followed. All such variations have to be ordered in writing, or if given orally, must be confirmed in writing. The ICE conditions require the contractor to undertake such ordered variations and, in general, they are to be paid for at bill rates or rates based on them. In some instances this may seem harsh on the contractor, since he may be doing work somewhat different from what he expected, and the rates so applied may seem to him too low. But the reverse can also happen, and some of the varied work may leave the contractor with a welcome extra profit, if the relevant rates happen to be set high at tender.

Extensive variations can make the contractor's task of constructing the works to his original programme impossible and can seriously affect his costs. They should be avoided if at all possible, but if they occur, the added costs can be taken into account by allowing for them in the rates set under the variation order. But ordered variations must not be so large as to alter the nature of a contract. This problem more usually arises when the employer decides to delete some substantial part of the contract works, such as a complete structure or a length of pipeline. A large deletion may so change the content of the contract that it may have to be re-negotiated, or maybe some agreement has to be reached to reimburse the contractor part or all of his intended profit on the deleted work. Clearly this is a matter for agreement between the employer and the contractor, and could not be ordered as a variation.

Under the **ECC conditions** (see Section 4.2(f)) the project manager may instruct a change to the works information and this has the same effect as a variation. The effect of such an instruction in the terms of ECC is to create a compensation event; one of the many such events listed in core Clause 60.1. On giving the instruction, the project manager asks for a quotation from the contractor, which is to include both proposed changes to the price and the time for completion. If the project manager does not accept the quotation he may ask for it to be revised or can make his own assessment of the effect of the instruction. The means of assessment depend on which of the options for payment has been selected but can include use of items in any bill of quantities.

Under **lump sum contracts** the ability to order variations may be much restricted, and may sometimes only be possible by pre-agreement. Normally there will be some contingency money in the contract, which the engineer is authorized to expend on necessary variations. Since there are seldom any unit rates in a lump sum contract, the engineer may have to request a quotation from the contractor for a proposed extra before he orders it. It depends on the contract provisions how he deals with a quotation which he thinks is too high. Sometimes he will have no power other than to negotiate a lower price from the contractor. If that fails he either orders the extra at the contractor's price or does not order it. If it is a matter of some importance he may decide to consult the employer on the matter. On lump sum or turnkey projects, care has to be taken to ensure that any extra required is a true addition, not included in or implied by the overall requirements of the contract. As may be imagined this is a fruitful cause of dispute.

On some lump sum contracts, while all above-ground work is paid for by means of lump sums, a small bill of quantities may be included for below-ground, that is foundation work, so that it can be paid for according to the prices entered by the contractor and the measure of below-ground work required. This covers the case where the extent of the foundation work may not be exactly foreseeable. Other lump sum contracts may include a schedule of rates to be used for pricing ordered variations, typically adopted in the case of electrical or plumbing contracts where additions of a standard nature are often found necessary.

17.4 Rates for ordered variations

Under the ICE conditions Clause 52 a proposed variation can be the subject of a quotation from the contractor either before or after it is instructed. If a quotation is not accepted an ordered variation can be valued in one of three ways:

- 'where the work is of similar character and carried out under similar conditions to work priced in the bill of quantities it shall be valued at such rates and prices contained therein as may be applicable'; or
- if not 'the rates and prices in the bill ... shall be used as the basis for valuation so far as may be reasonable failing which a fair valuation shall be made'; or
- the engineer can order the work to be carried out with payment to be made by dayworks if he thinks this necessary or desirable.

In addition, if the effect of any variation is such that any rate in the contract is 'rendered unreasonable or inapplicable' the engineer can fix such rate as he thinks 'reasonable and proper'. This allows the engineer to look at the effect of any variation on the contract as a whole and to allow modification of other rates if necessary. Hence if a variation has the effect of extending the time to complete the works, any time-related or similar preliminary rates can be adjusted to allow for the consequence of instructing the change.

Where bill rates do not directly apply, an appropriate bill rate can sometimes be deduced by extrapolation of quoted rates. For example, if rates exist for trench excavation not exceeding 1.5 m depth and not exceeding 2.0 m depth, a rate can be extrapolated for not exceeding 2.5 m depth. However, this simple approach is not always possible because rates often exhibit discontinuities. Thus if there are rates for 100 and 300 mm diameter pipelines, the rate for a 200 mm pipeline may not lie halfway between them. To fix a new rate it may be necessary to break it down into its component parts. The price of the pipe and its weight for handling and laying may not be pro rata to diameter because of increased wall thickness; and the trench excavation width may be virtually the same for a 200 mm pipe as for a 100 mm pipe. A rate derived from build-up of prices should be compared with one derived by deduction or addition from bill rates. The latter may prove fairer to both parties because bill rates will include the addition chosen by the contractor for overheads, risks and profit.

A problem arises when a bill rate or price which could be used for extra work appears unjustifiably high or low, either by error or, in the case of a high rate, perhaps by intention. One party or the other may feel it is unjust to use such rates or base new rates on them for varied work. However, it can be pointed out that the use of existing rates is what the contract requires, and other rates in the contract must have been correspondingly low (or high) to arrive at the tender total. Even when existing rates cannot be used directly, using them as a basis for new rates, or adopting similar levels of overhead and profit can be seen to give a fair result under the contract.

The problem of setting rates for new work or for omitted items, or where a quantity change of itself justifies a new rate, can sometimes prove difficult. The principle in these cases is, however, the same – the billed rates act as the predominant guide when developing varied rates, because they are the basis of contract. If this principle is departed from, it can be seen that many complications could arise in setting new rates since, if one bill rate is not adopted because it appears too high (or low), then either party could maintain the same applied to other bill rates, and there would be no clear basis for setting new rates.

It should be noted that the phrase 'Variation Order' is not used in most conditions of contract. Variations in the works are **instructed** (ICE conditions Clause 51) and **valued** (Clause 52). A 'Variation Order' then results as a record of the instruction and valuation.

In the United States the term 'change order' is used in lieu of variation order.

17.5 Variations proposed by the contractor

The contractor normally has no right to vary the works and the terms of the contract will specifically preclude this. But he can make suggestions as to how the work might be varied, for his own benefit or the benefit of the employer or both. He has no power to adopt his own suggestion; but if, say, he is unable to purchase an item required but finds an adequate substitute the engineer would no doubt agree. On occasion a good contractor will point out a change of design that has advantages, and the engineer should consider this because the knowledge of the contractor can assist in promoting a sound construction or reduced cost (see Section 17.1).

Situations can arise where the contractor's work does not accord with the stated requirements. This may be by default when materials or equipment have been ordered and delivered only for it to be discovered that they are not in compliance with the specification. Or it may be that workmanship is found unsatisfactory only after some work has been built, such as concrete of too low a strength having been used in part of the structure. Under the contract the engineer has no option but to reject the work; but it may be to the advantage of progressing the works and preventing delay if the engineer discusses with the employer and contractor the possibility of accepting what has been provided, but at an adjusted price. Clearly this is not possible if the difference means the works will be unsafe or not usable for their intended purpose, but the employer may be able to accept a lower quality finish or the possibility of increased future maintenance if the cost of the works is reduced.

Any substitutions offered by the contractor should be referred by the resident engineer to the engineer, who will decide if the employer's views should be sought. The employer is entitled to receive what was shown on the drawings and specified, and not something else. If any such change is to be accepted the full terms of agreement including price and time effects must be recorded in writing to avoid later arguments.

17.6 Claims from the contractor

The term 'claims' is loosely used and has several meanings, which can cause confusion unless the context within which the word is used makes the meaning clear. In ordinary parlance the word is used to mean 'claims for more money by a contractor which may or may not be payable', that is, for matters other than those for which payment is specified in the contract. However, most contracts formally recognize and define some types of 'claim' a contractor can submit.

Clause 53 of the ICE conditions sets out the procedure to be followed by the contractor if he wants to claim (a) a higher rate or price than the engineer has set under a variation order or in relation to some altered quantity under a bill item; or (b) additional payment he considers he is entitled to under any other provision of the contract. Under (a) the contractor must give notice of his intention to make a claim within 28 days of being notified of the engineer's fixing of a price. Under (b) the contractor must give notice 'as soon as may be reasonable and in any event within 28 days after the happening of the events giving rise to the claim'. The provisions with respect to (b) primarily relate to claims for encountering 'unforeseen conditions' or claims for delay. Both these are complex matters, which are dealt with separately in Sections 17.8–17.10.

Claims that arise concerning a rate or price set by the engineer for some varied work or excess quantity measured are often uncomplicated. Sometimes the facts need unravelling, such as – what activities is the rate to include; why do records of time or quantity spent on the operation differ between contractor and resident engineer? These matters have to be gone into in detail. The contractor may contend that the rate should allow for standing time, 'disruption' and 'uneconomic working'. There is truth in a contractor's claim that any rate set should allow for these matters. 'Uneconomic working' depends on the nature and quantity of extra work ordered. To order something additional to a contractor's current work can put him to considerable re-organization. For instance to order tie-backs to sheet steel piles after they have been driven involves obtaining extra steel, making extra excavation, and probably hiring welders. To get this organized may take some days, during which the contractor may not be able to start the next major operation scheduled on his programme.

Fairly frequent claims consist of the contractor claiming he should be paid for something for which there is no obvious measurement in the contract. This type of claim depends on whether the terms of the contract allow payment or not. This is when the specification, bills of quantities and method of measurement come under close scrutiny, because any inconsistency between them is liable to give the contractor at least some kind of case for payment. When specifications and bills of quantities are very large, the odd error will invariably occur. The resident engineer should take care not to agree an extra prematurely, he should check the contract first, to ensure that it does not include the extra in some bill rate.

Some contractors use confirmation of verbal instructions (CVIs) as 'claims' (see Section 13.3). The resident engineer or one of his staff tells the contractor that the blinding concrete looks too thin in places and must, as specified, be a minimum of 100 mm thick, and within an hour or so the resident engineer receives a signed CVI stating 'We are to thicken up blinding concrete at so-and-so'. At the bottom of the CVI form is printed 'and any extra work arising from the above instruction will be charged'. This sort of spurious claim has, of course, to be rejected immediately in writing by the resident engineer. Otherwise, if not contradicted and left on file, it may later be re-submitted by the contractor as a justifiable 'claim' long after the nature of the incident has been forgotten. If the resident engineer has, however, given a verbal instruction that justifies a claim for extra payment, he should confirm it in writing with precise details, and require the contractor to submit a detailed account of his costs promptly. He should also keep his own records of the work done by the contractor in response to the instruction.

17.7 Sheets submitted 'for record purposes only'

When a contractor considers some work entitles him to extra payment but the engineer does not immediately agree, the contractor may suggest that he should submit daywork sheets for it 'for record purposes only' (FRPO sheets), so that the quantity of alleged extra work can be agreed. This suggestion may seem reasonable, but it can result in the contractor's submitting scores (or hundreds) of FRPO sheets for everything, which he thinks he could claim as an extra. He can work on the basis that the more sheets he puts in, the greater is his chance of getting some extra payment, and so he is not over-concerned as to their accuracy or validity. The resident engineer, however, may not have the staff to check so many sheets and may consider it a waste of time to check them if many appear obviously invalid claims.

Therefore, on the first occasion when the contractor suggests submitting FRPO sheets, the resident engineer should refer the proposal to the engineer since, once the principle is accepted for one matter, it may be difficult to prevent submission of FRPO sheets for other matters. Under Clause 53 of the ICE conditions the contractor is required to give notice of a claim, and after that is required to submit full details of it. FRPO sheets are not recognized under ICE conditions, nor does the engineer have to evaluate a claim (if payable at all) on a dayworks basis if that is what is suggested (see Section 13.8). Hence the engineer may decide not to agree to submission of FRPO sheets and, if the contractor persists in sending them, he may advise the resident engineer not to reply to them, only to file them, putting notes thereon concerning their accuracy in case they later form the basis of a properly submitted claim. This avoids time-consuming correspondence and dispute on the sheets, which might inadvertently give the impression the contractor has a claim, which is valid in principle. If, however, dayworks sheets (whether labelled 'FRPO' or not) are submitted in support of some properly notified claim regarding extra work, the resident engineer must reply if he considers the sheets are invalid or incorrect, or may need to reject dayworks rates as a means of payment unless work has been instructed on that basis.

17.8 Clause 12 claims for unforeseen conditions

Among the more difficult and therefore more challenging types of claim are those relating to 'unforeseen conditions' – usually ground conditions. Clause 12 of the ICE conditions permits a contractor to claim extra payment:

> if the contractor encounters physical conditions (other than weather conditions or conditions due to weather conditions) or artificial obstructions which conditions or obstructions could not in his opinion reasonably have been foreseen by an experienced contractor.

There has frequently been criticism of this Clause 12 definition, but it has stood the test of many contracts over the years, and no alternative phrase has ever been put forward that works distinctly better. Some employers have tried deleting the provisions of Clause 12 entirely; but the contractor then adds a premium to his prices for the added risk he takes, so the employer pays this whether or not any unforeseen conditions arise. A point to be borne in mind if Clause 12 is deleted, is that it is usually impracticable to allow each tenderer to conduct his own site investigations, so he has no way of limiting his risk other than by raising his price. On a pipeline, for instance, the road authorities and private landowners would not permit each tenderer to sink his own test borings all along the route; nor may the employer allow each tenderer to sink test borings on the site of some proposed works.

A different attempt to avoid the problem of unforeseeable ground conditions is to specify the nature of the ground to be excavated as inclusive of practically everything, for example, in soft or hard material including gravel, cobbles, boulders, rock or concrete, running sand, etc. But if Clause 12 is left in the contract it over-rides such a specification because the extent to which any of these materials occurs remains undefined, so 'unforeseeable conditions' could still occur.

Although there is plenty of scope for the contractor to claim that things have not turned out as he expected, the criterion is whether 'an experienced contractor' could have foreseen the 'event' or not. To decide this with respect to ground conditions depends on the geotechnical information made available to tenderers together with any information readily available, such as that relating to the geology and soils of the area, and common experience locally. It needs to be remembered that when the contractor undertakes the obligation to construct the works he should have looked into these matters. Often it is not so much the event as such which is unforeseen, but its magnitude.

For example, test borings may show that hard bands of siltstone are likely to be encountered in tunnelling. But if, instead of occasionally appearing in the tunnel face and disappearing, a band manages to stay exactly in the soffit of the tunnel for a considerable length – this has occurred – this greatly adds to driving costs. The problem is that borings often reveal a range of ground conditions, but unless numerous borings are taken, they seldom disclose the degree of persistence and exact location of one particularly difficult condition. In fact, if an experienced resident engineer and the experienced engineer find themselves surprised by the 'unforeseen event' it is difficult to maintain that the contractor should have foreseen it. The problem has to be solved on the basis of reasonableness. A contractor could not reasonably be expected to foresee ground as uniformly bad when trial borings only show it to be of variable quality, good and bad.

The advantage of Clause 12 is that it permits many unforeseen conditions to be dealt with efficiently by a contractor with no dispute or problems of payment arising. It offers fair payment to a contractor so he will co-operate with the engineer in dealing with the conditions as effectively and economically as possible. Thus the employer pays only that which is necessary for dealing with the unexpected problem. Quite often the employer has to pay no more than he would have done had the condition been known beforehand and written into the contract. Thus both employer and contractor are fairly dealt with if Clause 12 is properly interpreted.

The **ECC conditions**, (see Section 4.2(f)) include for unforeseen physical conditions on a similar basis, classifying it as 'a compensation event' (Clause 60.1(12)). The test is worded, however, slightly differently from the ICE conditions, being conditions:

> which an experienced contractor would have judged at the Contract Date to have such a small chance of occurring that it would have been unreasonable for him to have allowed for them.

The effect may be much the same as for the ICE wording but has not yet been tested to the same extent by the courts. The outcome for an employer may, however, differ as the boundary between what is covered by the contractor's prices and what is not, may have altered by the difference in definitions.

17.9 Payment for unforeseen conditions

A problem arising with Clause 12 claims is assessing the cost of overcoming the unforeseen conditions. When the contractor has notified a claim under Clause 12(2) he has to give details 'as soon as practicable' of how he is overcoming or intends to overcome the unforeseen conditions, with an estimate of the cost and delay they will involve (12(3)). The engineer can step in and instruct the contractor what to do (12(4)). Since the contractor has notified he is making a claim, the provisions of Clause 53 also apply, which require the contractor to keep records of his work in connection with his claim, and send 'a first interim account' giving particulars of the amount claimed to date, followed by further accounts at intervals required by the engineer. The contractor is entitled to receive his costs 'reasonably incurred' in overcoming unforeseen conditions, 'together with a reasonable percentage thereto in respect of profit'.

In this first interim account the contractor will no doubt wish to include the cost of dealing with the unforeseen conditions some days before actually submitting his Clause 12 claim, and this is not unreasonable. 'Unforeseen conditions' do not always happen suddenly, they can be conditions which worsen gradually, causing the contractor increasing difficulty and delay as he tries to deal with the situation, until he realizes he has the basis for a Clause 12 claim. Until then the contractor may have no special records for that part of the work. So his 'first interim account' under Clause 53 may be sparse on detail but contain a large sum for work on the unforseen conditions before he sent in his claim.

While only the engineer can finally decide what should be paid, the work of finding out all costs 'reasonably incurred' will fall upon the resident engineer. Using his own records and the contractor's, he should endeavour to find data, which supports the contractor's claim.

Direct costs in meeting unforeseen conditions will comprise labour, materials and plant used. These should be ascertainable from records or, where these are not detailed enough, by judging what must reasonably have occurred, such as labour standing time or uneconomic working as problems created by the unforeseen conditions cause delay or the need to get more plant. Materials costs may have to include abortive temporary measures, such as timbering or temporary concreting. Plant costs may have to be divided into plant working and plant standing.

Indirect site costs comprise salaries and allowances for site staff, transport, office costs, plant maintenance, services and 'consumables' of all kinds that it may be impracticable to consider in detail. The contractor may be able to show what ratio they bear to gross labour costs; if they can only be shown in total to date, the current ratio has to be estimated, since the ratio will be higher in the early stages of the contract when site offices and services, etc. have to be set up, than later when more productive work is being undertaken.

Overhead costs for head office management and profit – usually expressed as a percentage on – have to be justified by the contractor showing they are in line with his usual practice.

Where some costs are difficult to elucidate due to lack of records, there are various estimating books published annually which can provide guidance, or be used as a check on the contractor's submissions.

When the unforeseen conditions occur and the contractor notifies his intention to make a claim in consequence, the engineer should report the matter to the employer. He should take into account the employer's views as well as the contractor's when considering whether the conditions could 'not reasonably have been foreseen by an experienced contractor'. The employer may also require to be consulted on the contractor's claims and take part in meetings with the contractor about them. Under the ICE conditions the decisions finally rest with the engineer, but he should endeavour to get agreement beforehand to his proposed decision from both contractor and employer, or at least their understanding of the reasons for his decision.

17.10 Delay claims

The handling of delay claims often poses difficulties. Under the ICE conditions Clause 44, the engineer can give an extension of time for the contract completion period if the contractor is caused unavoidable delay. The causes of delay can be numerous, including failure of the employer to give access to the site, or failure of the engineer to supply drawings as requested, or to approve the contractor's proposed methods of construction in reasonable time, etc. But the principal causes of delay are often variation orders for extra work and the incidence of unforeseen conditions (i.e. Clause 12 claims). In addition, Clause 44 permits an extension of the contract period on account of 'exceptional adverse weather conditions'. Clause 44 sets out the procedure to be followed which requires the contractor to give notice of the delay within 28 days of first experiencing the cause of delay or 'as soon thereafter as is reasonable'. The contractor has to give 'full and detailed particulars in justification of the period of extension claimed'. Although claims for delay are usually based on extra work ordered or caused by unforeseen conditions, there can be other delays not associated with extra work, such as when the engineer instructs the contractor to delay starting some foundation construction because of the need to conduct foundation tests.

Although a delay of some kind can be caused to a contractor's work, the delay of itself does not necessarily entitle the contractor to an extension of the contract period. The latter is a separate issue, which poses two particular difficulties:

- how to estimate the delay to the job as a whole caused by delay on just one operation (or a group of operations) when several hundred other operations are required to complete the job;
- how to estimate what extra cost, if any, is caused by the delay, over and above that which the contractor is paid for the extra work (if any) causing the delay.

The answer to the first question is illuminated by considering how critical path programming (mentioned in Section 14.5) would deal with a delay. Under that programming, if an activity lying on the critical path has its duration extended, then the delay to the whole job is likely to be equal to the activity delay. But, if the activity does not lie on the critical path, its increased duration can either have no effect on the time to complete the whole job, or it may create a new critical path, which is longer by some amount than the previous critical path. However, if it is possible to alter the sequence in which activities are undertaken, a further critical path may emerge which may be no longer than the original one.

The engineer has to consider whether the contractor could reasonably avoid delay to the whole project by undertaking other work available; hence, mitigating the delay. Thus if the contract comprises the construction of a single building for which the ground conditions turn out so unexpectedly bad that piling of the foundations has to be added, this would justify an extension to the contract period. The view cannot be taken that construction of the building could be speeded up to compensate, because this could involve the contractor adopting different methods for construction than he planned in his programme and might involve him in more cost. On the other hand, if there are several buildings, which the contractor has programmed to construct in sequence, and one of them is delayed by foundation problems, the contractor can divert his workforce to those not delayed, so there may be no need for an extension of the contract period.

The resident engineer's records are vitally important when considering delay claims. It is reasonable to allow a contractor some costs of disruption when he has to change unexpectedly from one operation to another, but it is unreasonable for him to leave his men doing nothing when there is work to get on with. Also a contractor cannot allege he is delayed by 'late receipt of engineer's drawings or instructions' when he is in no position to do the work because he is behind his programme, or his plant is broken down.

With respect to 'exceptional adverse weather' as a cause of unavoidable delay, in the UK this usually means wet weather, including flooding, holding up crucial earthwork constructions, such as embanking and road construction. A contractor normally allows about 10 per cent time for 'lost time' due to weather in the UK, but this depends on the nature of the works to be constructed. It should be noted that, under ICE conditions, an extension of the contract period on account of exceptional adverse weather, does not entitle a contractor to extra payment on account of the delay; though if there are items in the bill of quantities payable per week or month, such as for the maintenance of the resident engineer's offices, these would continue to be payable for the extended contract period.

The **ECC conditions** deal with weather by comparing actual weather conditions experienced on site, with the weather data supplied and set out in the contract. A 'compensation event' is then established if the weather conditions experienced can be shown to have a frequency of less than once in 10 years, that is, 1 in x years where x is greater than 10. As with all compensation events under the ECC this may lead to adjustment of both prices and time for completion.

17.11 Estimating delay costs

The cost which has to be evaluated due to delay to a contractor varies according to whether or not the delay justifies an extension of the contract completion period. If the delay does not justify an extension of the completion period, then the basic delay costs comprise such matters as standing time, lost time, and 'uneconomic working' for labour and plant. These can occur when the contractor has to stop work waiting for instructions, re-organize his work to cope with unforeseen conditions, or having to move labour and plant onto some other work available or as directed by the engineer. The 'lost time' by men and machines can be identified and costed, on a similar basis to that set out in Section 17.9.

Some other costs may have to be added, such as continuing to keep an excavation dewatered, or prolonged use of timbering to keep an excavation open, hired plant having to be retained on site longer, etc. These delay costs are separate from and additional to the rates set to cover the work actually undertaken. The latter rates should allow for the further difficulties and costs encountered as the work proceeds, such as continued de-watering for example.

If a delay justifies extension of the contract completion period, or it extends a major activity, then clearly some of the site resources have to continue for that much longer. Hence site on-costs must be added (see Section 17.9). On large jobs with several sub-agents or teams of staff, each may have to be considered separately to identify the effect (if any), which a delay has had in keeping them on site. Head office on-costs may also need to be added, but these are often a source of much confusion in relation to extension of the contract period. A contractor is not entitled to maintenance of a steady income from a site irrespective of what is actually happening at the time. An attempt should be made to identify any actual head office costs associated with an extension of the contract period using time sheets or other means; but if this proves impossible a general percentage addition which represents a reasonable proportion of head office costs to turnover costs can be added. Formulae such as Hudson or Emden may be proposed by contractors but it is important to recognize that these have no connection with any actual costs incurred.

It must be emphasized that it is the conditions of contract, which set out the delays for which payment of costs may be recovered by the contractor, and reference must always be made to them. The conditions will also generally define what is meant by 'cost', which usually excludes profit, although profit is deemed to be included in bill rates, which are often used for pricing variations. The ICE conditions specifically allow for profit when unforeseen conditions have to be dealt with.

The **ECC conditions** do not define cost as such (other than as a valuing mechanism) and may allow addition of profit to all price changes by means of the added fee. Both ICE and ECC conditions refer to financing charges as part of cost and these must be distinguished from interest on late payments (see Section 17.13). In this sense a financing charge is part of the cost itself and not due to any lateness. An example of a valid financing charge would be the cost of financing a retention deduction for longer due to delay. Claims for payment under ECC may not distinguish clearly between such financing costs, which are allowable under the contract, and claims for interest for a period following the end of the delay, which may not be allowable.

17.12 Quotations from a contractor for undertaking variations

On being instructed to vary the work or experiencing unforeseen conditions, it is sometimes the practice of a contractor to submit a quotation for dealing with the extra work involved, including perhaps some unspecified sum for overcoming any consequent delay alleged to have been caused. The resident engineer should not accept such a quotation, but should refer it to the engineer. The ICE conditions recognize that a quotation may be desirable and allow the engineer to request this for ordered variations. If the engineer finds the contractor's quotation unacceptable he can assess the variation at bill rates or similar (see Section 17.4). It is thus clear that the base line for any agreement is to be the existing bill rates. The position is different under design and construct contracts and the ECC whose provisions are discussed at the end of this section.

It may seem that acceptance of a contractor's quotation has the advantage that it avoids complicated problems of checking costs and assessing any delay. But it can prove highly advantageous to the contractor and disadvantageous to the employer. The contractor need not justify the amount of his quotation and, as he makes the quotation before he undertakes the necessary work, the engineer can only make an estimate of what the contractor's costs might be to check the quotation. Similarly, with no work done, there is no factual evidence as to what delay, if any, the extra work would cause. Sometimes work on two or more variations can take place at the same time, or otherwise be so closely connected (using the same equipment, for example) that it becomes difficult, or even impossible, for the engineer to judge whether the contractor's quotations contain elements of double charging of costs or double claiming for delay.

It should be noted that both the ICE design and construct conditions of contract and the ECC (see Section 4.2(d) and (f)) require the 'quotation approach' when variations are ordered. But those contracts are not necessarily based on a priced bill of quantities but often on lump sums, in which case the quotation method is appropriate (see Section 17.3). However, both the design and construct and the ECC are administered by the employer's representative or his project manager and not by an independent engineer, hence the contractor's power to quote in advance for extra work can be seen as strengthening his position. It is true that under those contracts the employer's representative or his project manager can reject the contractor's quotation, substituting his own, but this must inevitably raise a dispute.

17.13 Time limits and interest payable on late payments

The ICE conditions require the contractor to notify his intention to make a claim 'as soon as may be reasonable' but in any case within 28 days of meeting the unforeseeable conditions or cause of delay, or receiving notice of a rate set under a variation order. If the contractor is late in his notification of a claim, he may lose a right to that part of it, which the engineer cannot in consequence investigate properly (Clause 53(5)). When costs are ongoing, the engineer can require the contractor to submit further accounts of his claim at reasonable intervals. There is no specific time stated when the engineer must come to a decision concerning a contractor's claim, nor does he have to make a decision on a claim, which does not give 'full and detailed particulars of the amount claimed' (Clause 53(4) and (6)). However, once the engineer finds some payment due in respect of a claim he must certify this payment in the next interim payment to the contractor.

If the engineer decides that some payment is due to a contractor but unreasonably fails to certify it in the next certificate; or if the employer fails to pay part or all of an amount certified for payment by the engineer, the employer has to pay to the contractor 'interest compounded monthly for each day on which the payment is overdue or which should have been certified' (Clause 60(7)). The same clause provides that should a matter in dispute go to arbitration, and 'the arbitrator holds that any sum or additional sum should have been certified by a particular date' then interest will be payable on it, starting from 28 days after the engineer should have certified the sum. The rate stipulated is 2 per cent above the base lending rate of the bank specified in the Appendix to tender.

It should be noted that this interest is only applicable following a failure by the employer or the engineer. There is no general provision that interest is due for any gap in time between the costs being incurred and an amount being included in a certificate; much less so if any delay is due to a contractor failing to supply details of his claim.

17.14 Adjudication

The Housing Grants Construction and Regeneration Act, 1996, (see Section 1.6) introduced for the first time in English Law a requirement that all construction contracts must include certain terms. These include payment provisions as set out in Section 1.6 and also provisions to allow either party to take any dispute to adjudication at any time. Most standard conditions were immediately amended to comply with the law, but for those, which were not so amended, the provisions of *The Scheme for Construction Contracts Regulations 1998* will apply. The provisions of the Act apply to all construction works and work associated with construction in the UK such as architecture, design and surveying. There are some limited exceptions such as drilling for oil or gas, supply of materials and erection of machinery as set out in the Act but in effect most construction contracts and related consultancy agreements will be included.

The provisions of the Act require that should a dispute arise under the contract then this dispute can be referred to adjudication at any time and appointment of an adjudicator must follow within 7 days. The adjudicator must act impartially and is given wide powers to ascertain the facts and the law. He must reach his decision within 28 days or a longer period if agreed by both parties. This period is seen by some as too short for major and complex disputes but appears to have worked well in practice for straightforward matters including technical issues, simple claims and claims for non payment. While most standard forms of contract always had similar provisions, the introduction of adjudication has allowed many subcontractors easier and quicker access to an independent decision on matters affecting them directly.

17.15 Alternative dispute resolution

The time consuming and sometimes expensive traditional methods of resolution of disputes by reference to the courts or arbitration led to the adoption of processes known as **alternative dispute resolution** (ADR). These include direct discussion between executives of the parties; obtaining the advice of independent experts; or using a conciliator trying to find common ground, or of a mediator looking for an agreed solution. The ICE conditions permit either party to refer a dispute to conciliation procedure, provided the other has not already elected to go to arbitration. The difference between arbitration and conciliation needs to be appreciated. With arbitration each party states its case and is subject to cross-examination by the other party. The arbitrator's decision is based only on evidence submitted to him, although of course he can put queries to either party. But in a conciliation procedure the conciliator, often a professional engineer can investigate, and call for information on all matters he considers relevant to the dispute, and may interview the parties separately. This gives him a good chance of discovering the root cause of a dispute, enabling him to find a solution both parties can accept.

Of course for any method of conciliation or mediation to be successful, there must be a willingness in both parties to try to find a solution and the introduction of an outside independent party assists this process. Such methods of resolving problems are attractive due to reduced costs in employing lawyers and experts as well as in staff costs and in tying up senior management if they pursue arbitration or court action.

Many standard forms of contract refer to ADR methods and encourage the parties to try to settle disputes by such means. The introduction of provisions for adjudication into UK contracts has opened up the opportunities for early resolution of problems but there is still considerable interest in conciliation and mediation and the courts have encouraged parties to try such methods before commencing court actions. There is considerable debate concerning the relative merits of conciliation and adjudication. Conciliation proceedings are confidential, and the conciliator's recommendations cannot be quoted by either party in any subsequent arbitration. This aids reaching agreement as the disputants can state their views to the conciliator without prejudice. Adjudication is more formal. It is not a method of reaching agreement between the parties but a decision as to what the contract provides with respect to the matter in dispute. Any submissions to the adjudicator can be referred to in a subsequent arbitration, and the adjudicator may decide that he needs to employ specialist advice on technical or legal matters. Under adjudication the parties may feel it necessary to employ legal advice in presenting submissions and thus increase their potential costs.

17.16 Arbitration

Under the ICE conditions, Clause 66, either the employer or contractor can serve a Notice of Dissatisfaction on the engineer stating the nature of the matter objected to. The engineer has to give his decision on this matter within one month and, if either party is dissatisfied with it, they can issue a Notice of Dispute. The purpose of this procedure is to eliminate matters which can be resolved early by the engineer and to define any remaining dispute. A dispute thus established can be resolved by ADR, adjudication, or arbitration.

Although arbitration may seem a reasonable way of finally settling a dispute, it has disadvantages. It may take months to find an arbitrator both parties can agree to and who is willing and able to act. Both parties may decide to employ a lawyer to present their case because the dispute involves interpretation of the terms of the contract. Also one party may need to employ the engineer to act as witness, because the dispute also relates to the facts of the case, which lie in records kept by the engineer. If the employer is disputing the engineer's decision, this puts him and the engineer in a difficult position because the contractor may wish to employ the engineer as his witness, but the engineer may still be acting as engineer under the contract for ongoing work, and he can sometimes be an employee of the employer.

The lawyers who present the case for each disputant, although chosen for their experience of building contract disputes, may still fail to understand or make use of significant technical data having a bearing on the dispute. They may raise issues not previously in dispute, and use legal arguments concerning the contract, which the disputants feel are not relevant to the real matters of concern. Hence the outcome of an arbitration is uncertain and can be different from what either of the parties expected.

17.17 Minimizing claims and disputes

The key precautions which can be taken to minimize claims and disputes are:

(a) adequate site investigations;
(b) checking that the works designed satisfy the employer's needs and make reasonable provision for his possible future requirements;
(c) completing all design drawings, specifications, and arrangements for incorporation of equipment purchased separately, before tenders for construction are sought.

Under (b) the aim is to minimize changes during construction. Some may be unavoidable if the employer needs alterations due to some new regulation applying, or if his needs change due to something outside his control. But often his possible (but uncertain) changes of need can be catered for cheaply by incorporating a degree of flexibility in the design by providing extra space for additional equipment or runs of services.

Under (c) the construction contractor will see he has been given full details of what is required at the outset, reducing the likelihood of having to face additions and alterations during construction. This can encourage him to expect fast construction, which reducing his costs, enables him to submit his lowest price. Thus any extra time needed to ensure drawings and specifications are complete can be well spent.

To ensure designs are complete, the designers will need to ascertain the requirements of all specialist suppliers and contractors involved, such as the suppliers of mechanical and electrical plant, cladding, windows, cranes, lifts, heating and ventilating equipment. For major mechanical and electrical plant it is best for the employer to let separate contracts for their supply and erection, so they come under direct control of the designer as described in Sections 2.4(d) and 5.6.

18

Earthworks and pipelines

18.1 Excavating and earth-placing machinery

Bulldozers ('dozers') are used for cutting and grading work, for pushing scrapers to assist in their loading, stripping borrowpits, and for spreading and compacting fill. The larger sizes are powerful but are costly to run and maintain, so it is not economic for the contractor to keep one on site for the occasional job. Its principal full-time use is for cutting, or for spreading fill for earthworks in the specified layer thickness and compacting and bonding it to the previously compacted layer. It is the weight and vibration of the dozer that achieves compaction, so that a Caterpillar 'D8' 115 h.p. weighing about 15 t, or its equivalent, is the machine required; not a 'D6' weighing 7.5 t which is not half as effective in compaction. The dozer cannot shift material very far, it can only spread it locally.

A dozer with gripped tracks can climb a 1 in 2 slope, and may also climb a slope as steep as 1 in 1.5 provided the material of the slope gives adequate grip and is not composed of loose rounded cobbles. On such slopes of 1 in 1.5 or 1 in 2 the dozer must not turn, but must go straight up or down the slope, turning on flatter ground at the top and bottom. It is dangerous to work a dozer (and any kind of tractor) on sidelong ground, particularly if the ground is soft. Dozers cannot traverse metalled roads because of the damage this would cause, and they should not be permitted on finished formation surfaces. Sometimes a flat tracked dozer (i.e. with no grips to the tracks) can be used on a formation if the ground is suitable.

Motorized scrapers are the principal bulk excavation and earth-placing machines, used extensively on road construction or earth dam construction. Their movement needs to be planned so that they pick up material on a downgrade, their weight assisting in loading; if this cannot be managed or the ground is tough, they may need a dozer acting as a pusher when loading. This not only avoids the need for a more expensive higher powered scraper, but reduces the wear on its large balloon tyres which are expensive. The motorized scraper gives the lowest cost of excavation per cubic metre of any machine, but it needs a wide area to excavate or fill and only gentle gradients on its haul road. It cannot excavate hard bands or rock, or cut near-vertical sided excavations.

The **face shovel**, or 'digger' can give high outputs in most types of materials, including broken rock. It comes in all sizes from small to 'giant'; but for typical major excavation jobs (such as quarrying for fill) it would have a relatively large bucket of 2–5 m^3 capacity. The size adopted depends on what rate of excavation must be achieved, the capacity of dump trucks it feeds to cart away material, and the haul distance to tip or earthworks to be constructed. The face shovel would normally be sized to fill a dump truck in only a few cycles. The machine can only excavate material down to its standing level, and work a limited height of excavation face. Hence, if a deep excavation is required, the face shovel must 'bench in' and must leave an access slope for getting out when it has finished excavating. It must stand on firm level ground when working, and is not very mobile. It works in one location for as long as required, moving its position only as excavation proceeds. Its major advantage is its high output and ability to excavate in most materials.

The **hydraulic excavator** used as a hoe or backacter, cuts towards the machine. It is highly versatile. The larger sizes can cut to a depth of 6 or 7 m and excavate a face of the same height, slewing to load to trucks alongside. It can be used for lifting pipes into trenches, and 'bumping down' loose material in the base of a trench with the underside of its bucket. It can usually excavate trenches in all materials except rock; but sometimes has trouble in getting out hard bands of material that are horizontally bedded or which dip away from the machine. It can have a toothed bucket capable of breaking up a stony formation, or be fitted with a ripper tooth for soft rock or a hydraulic breaker for hard materials, or have a smooth edged bucket for trimming the base of a trench. A wide range of such machines are available, the smallest size often being used on small building sites; the larger sizes being used for large trench excavation and general excavation of all kinds.

The **dragline**'s principal use is on river dredging work from the bankside, and for other below water excavation. Although the machine is slow in operation and has a smaller rate of output than an equivalent hydraulic backhoe, it can have a long reach when equipped with a long jib and can excavate below its standing level. With a 15-m jib, it can throw its bucket 20–25 m out from the machine; hence its use for river bed excavation and bankside trimming. The dragline can also be operated to cut and grade an embankment slope below its standing level, or for dumping soil or rock on such a slope. A trained operator can be skilled at placing the bucket accurately to a desired position. The dragline offloads its material to dump trucks, but this tends to be a messy operation because the swing of the bucket on its suspension cable tends to scatter material.

The **wheeled loader** is widely used for face excavation in soft material, but its predominant use is for shifting heaps of loose spoil and loading them to lorries. It may have a bucket size of up to 5 m^3; it is very mobile and, being soft tyred, can traverse public roads.

The **grab** has a low output rate, but is used when sinking shafts in soft material, especially when sinking caissons kentledge fashion. It is also used occasionally for the job of keeping aggregate hoppers filled with concrete aggregates from stocks dumped by delivery lorries at ground level.

The **clamshell bucket** has a pincer movement, hydraulically operated, and is principally used for the construction of diaphragm walls. The bucket is fixed to a long rod which is lowered and raised down a frame held vertically (or at an angle) so that it can cut trenches up to 30 m deep in soft material, usually up to 0.6 m wide. The machine rotates so the clamshell can be emptied to a waiting dump truck.

Trenching machines can be used either for excavation of pipe trenches or construction of shallow diaphragm walls. They have a bucket chain cutter delivering material to the side of the trench or by additional conveyor belt can deliver to dump trucks. For hard ground the machine has special cutters cutting a groove at either side of the trench, with a third bucket cutter chain to remove the dumpling of material between.

18.2 Controlling excavation

The base of an excavation has usually to be trimmed level and cleared of disturbed or loose material so that it forms a solid base for concrete foundations, pipes or earthworks, etc. Specifications often call for the last 100 mm of excavation to be 'carried out by hand' – a costly procedure for the contractor which he usually seeks to avoid. The resident engineer is then faced with the problem of what alternative he will allow in lieu of hand excavation. In some types of ground, such as sandy or gravelly clay, it should be possible for the contractor to machine excavate to formation level if he uses a plain edged bucket to his machine, operates it with care, and uses the back of the bucket to re-compact any small amounts of loose material. Large open areas excavated by scraper or dozer have to be graded, and re-compacted using appropriate compaction machinery.

A formation in soft clay can be severely disrupted by tracked or wheeled excavating machinery. No amount of re-compaction of disturbed, over-wet clay will prove satisfactory; it has to dry out to a suitable moisture content before it can be rolled and compacted back. If a contractor uses a D8 to excavate down to formation level in such material, the formation surface will be so churned up by the grips of the D8 tracks that it will be rendered useless as a formation. If the contractor does not use the right method on soft clays, the resident engineer must warn him that all disturbed material will have to be removed and the excavation refilled with suitable other material or concrete at the contractor's expense. The excavation should be undertaken by using an hydraulic hoe working backwards so that it does not have to stand on the formation. As it works backwards, suitable hardcore or other blinding material can be dozed progressively forward onto the exposed formation and compacted. Alternatively it may be possible to use a flat tracked loader shovel to skim off the last 150–225 mm of excavation, any loose material being either removed by hand labour or rolled back with a light roller before placing of the base course for a road or blinding concrete.

The presence of springs in a soft formation material exacerbates formation finishing problems. Usually the specification will require spring water to be led away by grips or drains to a pump sump which is continuously dewatered to prevent softening of the formation. If springs are encountered and have not been anticipated, or the method of dealing with them is not specified or shown on the drawings, the resident engineer should report the situation to the engineer. Special measures are often required to deal with springs to ensure safety of the structure to be built on a formation containing them.

18.3 Haulage of excavated material

For large open excavations, such as when road cuttings have to be made and the material tipped to form embankments, or for building an earth dam from open borrow pit areas, the motorscraper is the most economical machine for excavating, transporting and placing clays and clay-sand mixes. But the gradients traversed need to be gentle and the motorscraper cannot pick up hard bands of material or rock, unless ripping beforehand can break up the material sufficiently. If hard or rocky material has to be excavated, the face shovel loading to dump trucks has to be used, the trucks commonly having a capacity of 50–60 t, sometimes larger. However, neither scrapers nor dump trucks can traverse public roads.

If the excavated material has to be routed off site via public roads to some dumping area, the excavated material has to be carted away by tipping lorries licensed for use on the public highway. Tipping lorries have a lesser capacity than dump trucks, usually in the range 10–30 t. A factor often having considerable influence when needing to transport material along public roads, is the reaction of the local road and public authorities who may object to the extra construction traffic and mud on the roads. If the local authority has also to give planning permission for dumping spoil on some given land, such permission may only be granted subject to restriction on the size of lorries used and their frequency of passage. This situation cannot be left for tenderers to find out; the employer has to obtain the necessary permissions and the contract must reproduce exactly the conditions laid down by the planning or other authority concerned and require the contractor to conform to them. If the restrictions limit the size and frequency of tipping lorries, the contractor may be forced to temporarily stockpile excavated material on site and double handle it in order to conform to his intended programme for construction and the haulage conditions laid down. This will raise his costs for excavation.

Assuming there are no planning restrictions, the contractor needs to choose that combination of excavating plant and haulage vehicles which achieves the required excavation rate at lowest cost. The face shovel or backhoe output must match the timing of empty vehicles back from the dumping ground and their loading capacity. This means that the excavator bucket size and loading cycle time must be such that one haulage truck is loaded and moving away by the time the next vehicle arrives. Hence, the cycle loading time for the excavator must be known. Thus if 10 m^3 haulage vehicles return at 5 min intervals, and the cycle loading time is 1.5 min, only three cycles of loading are possible so an excavator bucket size of 3.3 m^3 is required. Alternatively if the cycle loading time could be 1.25 min a 2.5 m^3 bucket would suffice. Allowance has to be made for the bulking factor and unit weight of the material to be excavated. The bulking of granular or soft material may range 1.1–1.3, through 1.4 for hard clays, to 1.6–1.7 for broken rock. Clays, clay–sand mixtures, gravels and sands may weigh 1.6–1.9 t/m^3 in situ while rock and hard materials may vary 1.9–2.6 t/m^3 in situ. The excavator bucket size has to allow for the bulking factor: for example, a 2 m^3 bucket may only lift and load 1.4 m^3 loose material at 1.4 bulking factor, so it will need seven loading cycles to fill a 10-m^3 tipper wagon. If this is too long a loading time for the required rate of output, an excavator with a larger capacity bucket is required.

Correct assessment of the bulking factor is financially important to the contractor, particularly in relation to the use of tipping lorries for offsite deposition of material. Whereas dump trucks used on site can be heaped, tipping lorries have a limited cubic capacity and payload, neither of which can be exceeded. Thus if a bulking factor of 1.2 applies, a 10 m^3 lorry will take away the equivalent of 8.3 m^3 net excavation; but if the bulking factor is 1.35 the 10 m^3 lorry will take away only 7.4 m^3 net excavation. If the contractor has based his price on the former but experiences the latter, he would find his price for disposal of material off site 12 per cent too low. This could mean no profit on the operation or a large financial loss, since there may be many thousands of cubic metres of material involved. In practice a contractor's past experience will guide him as to what plant to use, taking into account many other practical matters which apply, such as reliability of different types of plant, need for standby, margins for hold-ups, length and nature of haulage road, cost of transporting plant to and from the job, and hire rates for different sizes of excavator and haulage vehicles.

18.4 Placing and compacting fill

When the contractor assesses the amount of filling he will need to transport to achieve a given earthwork construction, he has to allow for:

- fill after compaction occupying more, or less, volume than it does in the borrowpit;
- settlement of the formation under the weight of new fill as placed;
- further compression of the fill after placement under the weight of the fill above.

Standards will be set in the specification for the permitted moisture content of the fill before it is compacted, and its density after compaction. For example, the specification may stipulate that fill type A must be compacted at a moisture content between 'optimum −1 per cent' and 'optimum +2 per cent'; while fill type B must be compacted at a moisture content between 'optimum +1 per cent' and 'optimum +3 per cent'. The optimum value is that determined by the standard compaction test, whether it is the 2.5 kg hammer method (the original 'Proctor' test) for embankments, or the 4.5 kg hammer method as used for roads. The density to be achieved will be specified as some percentage (e.g. 90 or 95 per cent) of the optimum under standard testing. Samples will have to be taken from the fill to find the optimum moisture content and density under the standard compaction test specified, and in situ density tests (see Section 12.11) must then be undertaken by the resident engineer to ensure the right density is achieved. Normally tests on fill materials will have taken place prior to the design of the earthwork and the results of these tests and the method of testing, etc. must form part of the data accompanying the specification. Care has to be taken to ensure that the method of specifying the required end result covers the range of materials likely to be encountered. It is then up to the contractor, from his experience, to know what type of plant he must use to compact the fill to the required standard.

Achieving the required moisture content may present difficulty. In wet weather the borrowpit material may be too wet to use and the formation may be too wet to work on. The resident engineer may have to instruct the contractor to cease working when such conditions occur. There is little that can be done to protect borrowpit material against excess rainfall. The formation can be partly protected against rainfall by rolling it to a fall with a smooth wheeled roller at the end of each day's placing. Sometimes an attempt to protect the formation by laying sheeting over it is adopted, but this is seldom practicable if the site is windy. If the material is too dry for placing it must be watered. Although watering at the borrowpit can be helpful it is usual to water-spray the spread material from water bowsers. Some mixing of the material by dozer may be necessary after watering to avoid only the surface material being wetted. In hot dry climates more than the theoretical amount of water may need to be added because of the high evaporation rate applying. A considerable amount of water may be needed, involving the use of more than one water bowser.

18.5 Watching fill quality

When fill from a borrowpit is of variable quality the resident engineer needs an inspector to watch the fill quality as placed, with power to reject unsuitable material or call in the resident engineer in cases of doubt. Although the borrowpit must be examined to point out to the contractor where suitable and unsuitable materials appear to exist, the actual watch on material quality must take place as it is dumped, spread and compacted. The characteristics of a material are then more clearly revealed. There is not time to conduct in situ density tests: the contractor has to know immediately whether he can continue placing the material. Familiarity with the behaviour of suitable material as it is compacted and trafficked will soon indicate its characteristic behaviour. Excess of granular material, for example, sand or gravel in clays is easily observed, while too much clay or silt in a clay-sand mix is evidenced by the behaviour of the material under traffic. 'Cushioning' or 'bounce' under the wheels of lorries passing across the formation are signs of inadequate compaction which may be due to the material being too wet or containing too much clay or silt. Severe rutting by lorries can indicate material too wet or too clayey. Change of colour of a clay, on the other hand, may not indicate any change of suitability. The contractor must be warned immediately when material being placed appears unsuitable. If the placing is stopped after a few loads of unsuitable material have been delivered, these can usually be 'lost' by dozing the material out to mix it with previously placed suitable material.

Purpose mixing of two different kinds of fill is seldom practicable. It may be difficult to ensure that loads of the two materials are delivered in the right ratio and, if they are not clearly distinguishable apart by appearance, the mixing may be haphazard and incomplete. If two dissimilar materials must be used, the designer should preferably devise some means of zoning each separately. When zoning is adopted, the resident engineer should check from time to time that a supposed difference between materials is occurring because material from a borrowpit can change its composition gradually.

In situ density tests need to be taken to prove compliance with the specification; but the sand replacement method as described in Section 12.11 takes some hours to complete – so it is a record of past achievement and cannot be used as an instant control measure. The moisture content can possibly be quickly measured by using an appropriate moisture meter, but judging by eye can be equally effective and has the advantage that the whole area of placing can be kept under survey. The compaction equipment used by the contractor will vary according to the nature of the fill. Apart from the use of a large dozer to spread, compact and vibrate fill in place; the passage of laden dump trucks across a formation achieves a substantial degree of compaction. Hence, the contractor will usually arrange a method of placing material that makes effective use of the compactive effort of the delivery vehicles.

18.6 Site roads

A contractor who pays insufficient attention to the right construction of site haulage roads runs the risk the road will begin to break up and cause delay just at some crucial time of construction, such as when autumn rains begin and the contractor is hoping to get filling finished before the heavier rainfall of winter occurs and delays construction. Pushing hardcore into the worst patches is no real solution, and more troubles come when haulage lorries get bogged down in the road or break a half shaft. For heavy construction traffic, a road must be thick enough; have deep drainage ditches either side; be made from good interlocking angular large material at the base and similar smaller material above; and be formed to a camber or crossfall which sheds rainwater. Poor construction is more liable to occur on flat ground where the temptation is strong not to dig out more than seems necessary, and not to dig deep enough side ditches to keep the road construction dry. But once the proper precautions are taken the road will stand up and need little more than re-grading and rolling from time to time to keep the surface in good condition and able to shed rainwater.

18.7 Trenching for pipelines

The hydraulic hoe or backacter is the machine most widely used for trench excavation for pipelines. In hard ground, rock or roads, the trenching machine might be used which has been described in Section 18.1. Depths for water and gas pipelines are usually the pipe diameter plus 1 m. For sewers, greater depths are often required to maintain falls. When flexible plastic pipes are used, especially in the smaller diameters, pipe joints can be made above ground, the pipe being snaked in. Bottoming of the trench can be achieved by using a straight-edged bucket without teeth, and the backhoe can also place soft material or concrete into a trench on which to bed pipes or fully surround them. Provided no men are allowed in the trench, timbering can thus be avoided. When large diameter steel pipes with welded joints have to be laid, a string of several pipes may be welded up alongside the trench, and dozers equipped with side lifting booms can lower the string of pipes into the prepared trench. This reduces the amount of timbering and excavation of joint holes necessary which need only be arranged where successive strings have to be jointed together.

The principal defects occurring on pipelines come from defective joints and pipe fracture due to settlement of a pipe on a hard band, large stone or lump of rock in the base of the trench. The use of the hydraulic hoe makes the preparation of an even bed for the pipe easier to achieve, especially on suitable selected soft granular fill. However, the base of the trench and the bedding along each length of pipe must be carefully boned in before the pipe is lowered to ensure each pipe is fully supported along its body.

For non-flexible pipes of ductile iron, asbestos cement, steel or concrete it will be necessary to joint them after laying. Sufficient access is then required for the jointer to make the joint properly, and support to the trench sides will be essential in every case where there is not absolute certainty there can be no slip of material into the trench. Falls of material into trenches are a major hazard in civil engineering, and adoption of a consistent, rigorously applied safety approach is the only way to prevent accidents. The damaging weight of even a small fall of earth must be borne in mind.

While it will be obvious that gravity sewers must be laid to a fall, it is sometimes not appreciated that pressurized trunk water mains should be laid to a minimum rise or fall. The preferred minimum gradients are 1:500 on a rising grade in the direction of flow; and 1:300 on a falling grade. The former would be to an air release valve, the latter from the air valve to a washout or hydrant. Thus the levels of ground ahead of the pipelaying must be prospected to locate suitable high and low points, and intermediate points where an increase or decrease of grade is necessary. The pipeline between such pre-determined points should follow an even grade. In flat ground it may not be possible to comply with the foregoing grades, but it is still advisable to give uniform rises to air valves and falls to washout positions. In built-up areas pipelines can generally follow the requisite cover below ground surface because branches and connections will release air, and hydrants will be used as washouts.

Backfill to pipes should always be of selected soft or fine granular material to 150 mm above the crown of the pipe. Few contractors in UK would fail to do this, but on some contracts overseas the resident engineer may need to stop the contractor from dozing the excavated hard material straight back into the trench irrespective of the rocks it contains which would at the least damage the sheathing to pipes.

18.8 Thrust blocks and testing pipelines

The resident engineer may have to ensure that thrust blocks to pipes are adequate. A block acts primarily to transfer pipe thrust to the natural ground against which the block abuts. Hence, the nature of the ground is important. The force to be resisted on a bend with push in joints is shown in Fig. 18.1. The internal water pressure taken should be the maximum static pressure occurring, plus an allowance for surge pressure. The block bearing area against the

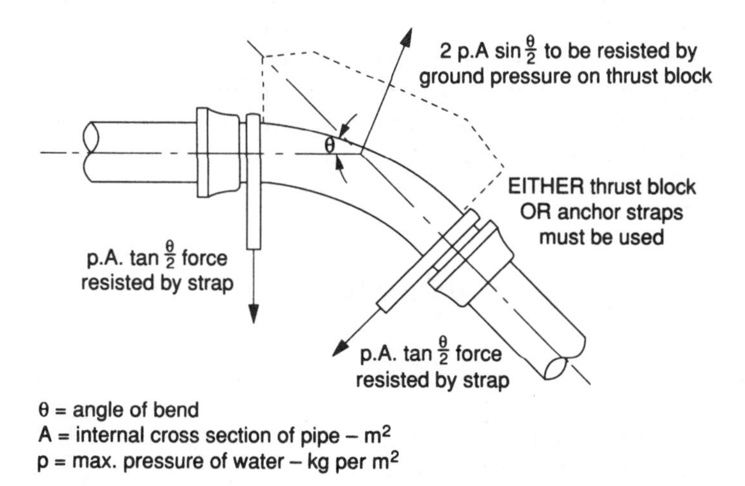

Fig. 18.1. Thrust on a pipe bend. The maximum pressure taken should be maximum static plus an allowance for surge

natural ground has to be sufficient to mobilize adequate resistance from the soil against pipe thrust. Where a bend points down, a weight block below the upper joint is usually necessary, the joint being strapped down to it. Particular care has to be taken when a bend down is required at the bank of a river or stream. There may then be very little ground resistance to prevent the bend blowing off if the joint is the usual push-in type, hence the upper joint of the bend may require tying back to a suitable thrust block to resist the hydraulic force tending to push the bend off.

The watertightness test for a pipeline should be prescribed in the specification. Care must be taken to fill the pipeline slowly to ensure release of all air. The usual practice is to fill from one end, having all washouts and hydrants open. The latter are progressively closed along the line when they cease to emit air or mixed air and water. This may take a considerable time, as air pockets may get trapped and only slowly disperse. A satisfactory test cannot be achieved if an air pocket is left in the pipeline. After filling, the pipe should stand under pressure for 24 h before testing. This permits pipe expansion and absorption of water by asbestos cement or concrete pipes or by any mortar lining of steel or iron pipes. Normally pipelines are tested between valves, but on a trunk main these may be so far apart that temporary stop ends may be needed to test the pipeline in reasonable lengths. Especial care is needed in testing partly completed pipelines to ensure they are properly restrained against the high test pressures.

An experienced pipelaying contractor will know that care in making joints on a pipeline is rewarded many times over when the pipeline test shows the line is satisfactory. It is rarely possible to leave joints exposed before testing, so if a test fails it may take much time and trouble to find the leak causing it. When finally found it may be as simple as a twisted rubber ring in a push fit joint. It is not easy to make clean perfect joints in a muddy trench. The contractor can save himself money if he gives the jointer every facility to make a good joint – easy access to see the underside of the pipe spigot, buckets of water and plentiful clean rags to ensure that joint faces and joint ring are scrupulously clean before the joint is assembled. No grease or jointing compounds should be permitted other than that which the pipe manufacturer recommends. Welders also need sufficient room and good lighting to make sure welds are adequate and should not be expected to weld up badly aligned pipes.

18.9 Handling and jointing large pipes and fittings

All ductile iron or steel pipes and fittings must be handled with proper wide lifting slings to prevent damage to their sheathing or coating. The use of chains or wire ropes 'blocked off' pipes with pieces of wood should not be permitted by the resident engineer. Apart from possible damage to the coating, the packing pieces may slip out when the chain or wire rope slackens and the pipe may fall. Handling of heavy pipes and fittings must be done with every precaution.

The crane handling a heavy pipe must not slacken off until the pipelaying foreman is certain it is safe to do so, and no man would be put in danger if the pipe should move. Timber props, packings and wedges in adequate numbers should be available to secure the pipe before it is finally moved into position for jointing by slow jacking or barring to get it into position, the wedges being continuously adjusted to keep the pipe from moving unexpectedly.

Where large diameter pipes and fittings, such as bends, have to be fitted together there is often difficulty in getting them set so that their joints match accurately, especially when a bend must be fixed at an angle to the horizontal. Before lowering such a bend into position it is worthwhile measuring it to find and mark, on the outside of the pipe at both ends, the diameters on the true axis of the bend and at 90 degrees to it. These should be accurately marked with a chiselled or indelible pencil line on white paint, not marked with chalk which will rub off. It is quite difficult to locate the axis of a bend accurately, even when the bend is above ground; and can be frustratingly difficult when the bend is laid in some trench or basement. It is also good practice to put a mark round the spigot end of a pipe showing how far it must be inserted into a socket. The relationship of this line to the socket face indicates whether the alignment is satisfactory. Spigot and socket pipes need to be lined up within 1 degree to achieve a good joint; pipe flanges have to be lined up exactly parallel, with the bolt holes exactly matching, and as close as possible after insertion of the necessary joint ring before final drawing together of the flanges by progressive tightening of the bolts. To set a 1.2 m diameter 45 degree bend accurately to join a horizontal pipe at one end and an inclined pipe at the other may take a gang of four men and a crane driver two or three hours. If things do not go well it may take much longer.

19

Site concreting and reinforcement

19.1 Development of concrete practice

Although many contractors now use 'ready-mix' concrete where it is convenient and economic, there are still many projects on which concrete is mixed on site. On remote sites and sites overseas there may be no ready-mix suppliers. If large pours are required it may be more economic for a contractor to produce concrete on site. In other cases a contractor may fear a ready-mix supplier would not be able to cope with variations in his site requirements. Traffic holdups can cause delay to delivery lorries, and create 'bunching' of deliveries. No contractor likes to see a partially completed pour of concrete moving towards its initial set when no further ready-mix lorries arrive. Concrete produced on site is under his control, he can start concreting as soon as formwork is ready, and can stop concreting in an organized manner if some difficulty arises in placing. Such problems are the contractor's but if he proposes to use ready-mix, the resident engineer should check that the supplier can meet the specified requirements for concrete. Some ready-mix suppliers, for instance, may not supply a 40 mm (1.5 in.) size coarse aggregate mix.

On the majority of construction sites the concrete used is made of natural aggregates such as gravel, sand and crushed rock mixed with water and ordinary Portland cement to BS12:1978.

For many years, concrete mixes were used as specified in BS CP114:1948, *The structural use of reinforced concrete in buildings*. This contained a table of recommended standard concrete mixes by weight, which is reproduced in Table 19.1 because it sets out familiar mixes still widely used for unsophisticated concrete work. In 1981, BS 5328, *Methods for specifying concrete* was produced. This was later re-issued in four parts – BS 5328:1997 Part 1, *Guide to specifying concrete*; Part 2, *Methods for specifying concrete mixes*. (Parts 3 and 4 were produced in 1990, the former dealing with producing and transporting concrete, and the latter with sampling and testing concrete.)

Table 19.1

British Standard CP114:1948 Standard mixes by weight

Equivalent volume mix proportions cement:sand: aggregate	*Works cube at 28 days (N/mm²)*	*Dry weight of aggregates per 50 kg cement (kg)*				
		Sand	*Coarse aggregates*			
			38 mm Max. size		*19 mm Max. size*	
			Workability		*Workability*	
			Low	*Medium*	*Low*	*Medium*
1:1:2	30	65	165	135	145	110
1:1½:3	25.5	80	200	165	165	135
1:2:4	21	90	225	190	190	155
Compaction factor			0.82–0.88	0.88–0.94	0.82–0.88	0.88–0.94
Slump (mm)			25–50	50–100	12–25	25–50

Notes: (1) Weights are based on use of Zone 2 sand (see Table 19.4) aggregates of relative density 2.6. If Zone 3 sand (finer) is used, reduce sand weights by about 10 kg and increase coarse aggregate by same amount. If the sand is crushed rock, reduce weight of coarse aggregate by 10 kg. (2) Columns for aggregate size have been reversed from that shown in BS 114 to agree with order used in Tables 19.2–19.4.

BS 5328 has been withdrawn in December 2003 and superseded by BS 8500:2002 Parts 1 and 2 which adopt the European Standard EN 206-1:2000 which is a complex, many-paged document covering many types of concrete and requirements for differing circumstances which do not need to be considered here. However, it maintains the same four categories of concrete specification as BS 5328:1997, namely – 'designed concrete'; 'prescribed concrete mixes', 'standard concrete mixes', and 'designated concrete'. These are described in Section 19.2.

Where a characteristic strength at 28 days is specified (as in 'designed concrete'), it is defined as a grade, for example, 'C.30' where 30 represents the nominal cube strength in N/mm^2. The strength is that which can be expected to be achieved with proper control of the quality of materials and the mixing. There is also an important grade of concrete specified in BS 8007:1987, *Design of concrete structures for retaining aqueous liquids*. This is particularly for concrete works associated with the water and wastewater industries, and is classed as grade C35A.

Since the production of CP 114:1948, a need for higher strength concretes, coupled with more detailed studies of the chemistry of concrete and the advantages of using different types of aggregates, cements and admixtures to meet varying conditions, has led to requiring concrete mixes to be designed to meet some specific strength together with such other requirements as are considered necessary. BS 8500 now sets out the main current requirements; while a number of other British Standards cover the use of special aggregates, special cements and additives. When any of these special ingredients are to be used, the procedures to be followed should be detailed in the specification or else provided by the engineer. In the material which follows only the ordinary matters which those on site will be expected to deal with are described.

It should be mentioned here that, in recent years there has been an increasing trend to use cement blends. These may be either Portland cement/pulverized fuel ash (PFA), OPC/PFA, or Portland cement/ground granulated blast furnace slag (GGBS), OPC/GGBS.

PFA is a residue of pulverized coal burnt in the furnaces of many modern power stations. It can be supplied to the concrete mixer as a component of a ready-blended cement, or as a separate material with its own storage and handling facilities. The advantages of the use of PFA in concrete are: overall economy of materials, improved workability and compactability, reduced water content, reduction of heat evolution, increased resistance to chemical attack (sulphate or acid). The advantage of improved workability is of considerable benefit where concrete pumping is required.

GGBS is a by-product of iron manufacture, where slag issuing from a blast furnace at a temperature of approximately 1500 degrees centigrade is rapidly quenched in water. This material is subsequently dried and ground to a fine powder which again can be supplied to the concrete mixer as a component of a ready-blended cement, for example, Portland blast furnace cement (PBFC), or as a separate material with its own storage and handling facilities. The advantages of the use of GGBS in concrete are: increased strength over the longer term (slower strength gain at first, then catching up and overtaking normal OPC concrete at 28 days and beyond), reduction in water content for equivalent cohesion, flow and compaction particularly when pumping, reduced heat evolution especially with thick sections and improved resistance to most forms of chemical attack (sulphate or acid). This is particularly advantageous in foundation works subject to sulphate attack.

19.2 Standards for concrete quality

The specification should define the mixes or grades of concrete required and in what parts of the works each mix or grade is to be used. As mentioned in the preceding section, both BS 5328-2:1997 and BS 8500:2002 describe four classes of concrete mixes – designed mixes, prescribed mixes, standard mixes, and designated mixes.

Designed mixes are specified by the purchaser stating the characteristic strength required, maximum size of aggregate and minimum cement content, leaving the supplier to design the mix proportions.

Prescribed mixes are specified by the purchaser stating the proportions of the mix constituents required – cement, aggregate, size and type, etc. – the purchaser being responsible for the performance of the mix.

Standard mixes are set out in a Section 4 Table 5 of BS 5328-2, and also in BS 8500 where they are called *Standardized Prescribed Concretes*. They are for concrete of characteristic strengths from 7.5 to 25 N/mm^2 and BS 5328 gives the

Table 19.2

Standard mixes as Table 5 of BS 5328-2:1997 also described as Standardized Prescribed Concrete in BS 85000:2002

Standard mix	*Characteristic compressive strength at 28 days (N/mm^2)*	*Constituents (Cement and Total aggregate) (kg)*	*Maximum size of aggregate* 40 mm		20 mm	
			Workability		*Workability*	
			Medium	*High*	*Medium*	*High*
ST1	7.5	Cement	180	200	210	230
		Aggregate	2010	1950	1940	1880
ST2	10	Cement	210	230	240	260
		Aggregate	1980	1920	1920	1860
ST3	15	Cement	240	260	270	300
		Aggregate	1950	1900	1890	1820
ST4	20	Cement	280	300	300	330
		Aggregate	1920	1860	1860	1800
ST5	25	Cement	320	340	340	370
		Aggregate	1890	1830	1830	1770
Workability – slump (mm)			50–100	80–170	25–75	65–135

mix proportions for size of aggregate, and workability. Table 19.2 reproduces Table 5 of BS 5328-2. The lower grades of characteristic strength 7.5, 10.0 and 15 N/mm^2 are intended for use in mass concrete filling to strip footings, blinding concrete, and similar.

Designated mixes are for mixes to meet special requirements, such as for sulphate resisting concrete, etc. and also for ready-mix concrete for which the supplier is required to hold a current product conformity certificate, to BS EN 150 9001.

The **characteristic strength** of a mix is defined as – 'that value of strength below which 5 per cent of ... strength measurements ... are expected to fall'. On a statistical basis, cube strength test results on a given mix are found to follow a 'normal distribution' that is 50 per cent of test results are above the mean X, and 50 per cent below. If only 5 per cent of results are to fall below a required value P, then the mean strength X must obviously be higher than P. From the characteristics of a normal distribution curve, the value of X has to be $1.64S$ higher than P, that is, $X = (P + 1.64S)$ to achieve not more than 5 per cent results below P, where S is the standard deviation[1] of the test results obtained. If only 2.5 per cent of results are to fall below P, then X must be $(P + 1.96S)$.

[1]The standard deviation, S, is defined by

$$S = \sqrt{\frac{(X - x_1)^2 + (X - x_2)^2 + \cdots + (X - x_n)^2}{n - 1}}$$

where there are n test results $x_1, x_2, \ldots, x_n$ and X is their mean value.

Reliable values of S cannot be obtained unless enough sample results are available. BS 5328 requires a minimum of 30 tests to be made before assessing S; and for $n = 30$ to 100 tests, the value S' should be taken where $S' = (0.86 + \sqrt{2/n})S$, with a minimum value of 6 N/mm^2. The DoE manual, *Design of normal concrete mixes* (1988), recommends that for Grades C20 and above, S should be taken as not less than 8 N/mm^2 until more than 20 test results are available, and a minimum value of 4 N/mm^2 should apply however many tests are taken. For trial mixes, values of $2S$ would also normally be taken (rather than $1.64S$) to give a further margin of safety. However, a less exacting approach is also allowable as described in the next section.

19.3 Practical compliance with concrete standards

It can be seen that although designing a trial mix to meet a given grade can be done within a reasonable time using recommended minimum S values, the proving of a mix by statistical analysis of cube strengths is a lengthy business. At least 20 batches of concrete would have to be made up (preferably 30 or more) and at least two cubes from each batch tested at 28 days. (BS 5328 requires the mean of two cubes to be taken in each case.) The statistical method is therefore mainly used to monitor concrete quality when large quantities are being placed, and is not practicable for small sites. However, BS 5328 laid down simpler criteria for concrete of grades C20 and above, namely:

- the average cube strength of four consecutive samples must exceed the desired characteristic strength by 3 N/mm^2;
- no single test shall be more than 3 N/mm^2 less than the characteristic strength.

Thus four batches tested are the minimum requirement for testing of a trial mix. As an example, a trial mix for Grade C25 characteristic strength would be satisfactory if samples from each of four batch mixes give an average strength of not less than 28 N/mm^2 with no single cube strength under 22 N/mm^2.

The specification may also set other requirements such as minimum cement content, minimum density of concrete, and minimum tensile strength under bending. BS 1881 (1981): *Methods of testing concrete* describes (inter alia) standard methods of sampling mixes, making and curing test specimens, density testing, and tensile testing – the latter labelled 'determination of flexural strength'. Cubes are usually tested in an off-site laboratory. A density test can be carried out on site provided it is defined as on 'fresh, fully compacted concrete' (see Section 19.8). The requirement for the flexural test is shown in Fig. 19.1. A test apparatus is simple to make on site, and the tensile strength of the concrete is easily calculated from the value of the central load which causes failure of the beam. This tensile strength is important in water retaining concrete, and can be helpful in deciding when support props can be removed from slabs and beams if the test beam is cured alongside and in the same manner as such slabs and beams. However, a margin of safety must be allowed for possible differences in compaction between the test beam and the in situ concrete.

To keep a watch on concrete quality during construction a record of the type shown in Fig. 19.2 can be used on which are plotted 7- and 28-day concrete cube strength test results. The figure illustrates a decline in strength at 7 January, remedied later by reducing the water/cement ratio.

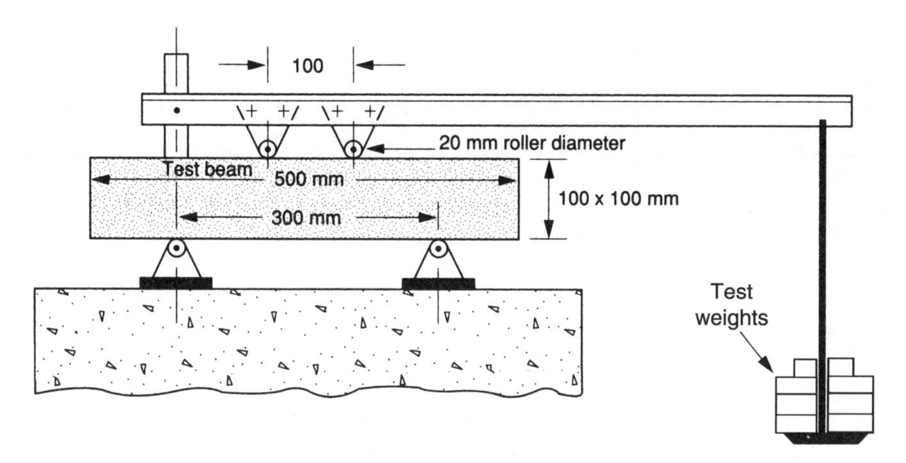

Fig. 19.1. Flexural test as BS 1881 on concrete beam for 20 mm maximum aggregate size. For simple site testing (see text) it may suffice to use a single central roller to apply the load

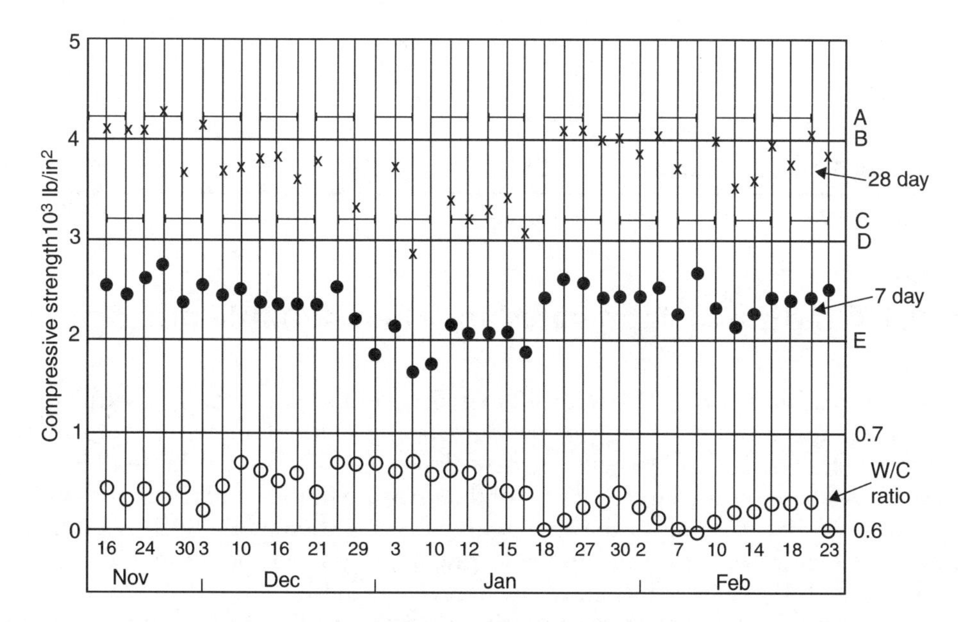

Fig. 19.2. Graphical check on concrete test results. A, upper action line 4200 lb/in^2 (29.0 N/mm^2); B, upper warning line 4000 lb/in^2 (27.6 N/mm^2); C, lower warning line 3200 lb/in^2 (22.0 N/mm^2); D, lower action line 3000 lb/in^2 (20.7 N/mm^2); E, 7 day warning line 2000 lb/in^2 (13.8 N/mm^2). Each point mean value of five test results.

Table 19.3
BS 5328:1981 Ordinary Prescribed Mixes*

Grade of concrete	*Weight of aggregate to be used per 100 kg of cement (kg)* Max. size 40 mm		Max. size 20 mm	
	Workability		*Workability*	
	Medium	*High*	*Medium*	*High*
C20P	660	600	600	530
C25P	560	510	510	460
C30P	510	460	460	400
Workability – slump (mm)	50–100	80–170	25–75	65–135

*Grades C20P and C25P were later designated Standard Mixes ST4 and ST5 in BS 5328:1997 as shown in Table 19.2.

Table 19.4
BS 5328:1981 Percentage by mass of fine aggregate to total aggregate

Grade of concrete	*Sand grade*	*Percentage of sand (fine aggregate in total aggregate)* Max. size 40 mm		Max. size 20 mm	
		Workability		*Workability*	
		Medium	*High*	*Medium*	*High*
	Zone 1	35%	40%	40%	45%
C20P	Zone 2	30%	35%	35%	40%
C25P	Zone 3	30%	30%	30%	35%
C30P	Zone 4	25%	25%	25%	30%
Workability – slump (mm)		50–100	80–170	25–75	65–135

Grades C7.5P, C10P and C15P have been omitted.

19.4 Grading of aggregates and their suitable mixing

The 1981 edition of BS 5328 (referred to in Section 19.1) provided a useful table showing the amount of aggregate per 100 kg cement in what were then termed (see Section 19.2), *Ordinary prescribed mixes* as shown in Table 19.3, and the percentage of fine aggregate to total aggregate in these mixes as shown in Table 19.4. The grading of the fine aggregate was then as BS 882:1973 which defined the grading for four **zones** of fine aggregate as shown in Table 19.5. Although BS 882:1973 has now been revised, Tables 19.3–19.5 are still of practical use as a guide to determining the ratio of fine to coarse aggregate required to make a dense mix. Of the four Zone gradings shown in BS 882:1973, Zones 2 and 3 were the most used for forming a suitable concrete mix. Zone 1 grading (the coarsest) tended to give a harsh concrete and also

Table 19.5

BS 882:1973 Grading of fine aggregate

BS410 sieve (mm)	*Percentage by weight passing BS sieve*			
	Zone 1	*Zone 2*	*Zone 3*	*Zone 4*
10.0	100	100	100	100
5.0	90–100	90–100	90–100	95–100
2.36	60–95	75–100	85–100	95–100
1.18	30–70	55–90	75–100	90–100
0.60	15–34	35–59	60–79	80–100
0.30	5–20	8–30	12–40	15–50
0.15	0–10	0–10	0–10	0–15

Note: Later versions of BS 882 have substituted Grades C (coarse), M (medium) and F (fine) for the above four zones.

was often not procurable; and Zone 4 was usually avoided if possible because it contained too much fine material for producing the best concrete. (The revised version of BS 882 in 1983 no longer defined four Zones for fine aggregates, but substituted 'Coarse', 'Medium' and 'Fine' gradings which are too wide in range to be of practical use for mix design purposes.)

In practice samples of the fine and coarse aggregates proposed to be used should be sieved to find their typical grading. Sometimes it is found that the coarse aggregate contains a substantial proportion of fines (below 0.5 mm), while the fine aggregate may frequently be of a uniform size. Consequently various ratios of coarse to fine aggregate must be tried out to see which gives the best mix. Envelopes of suitable grading curves for 20 and 40 mm maximum size aggregate are shown in Fig. 19.3. The first trial mix can adopt a ratio of fine to coarse aggregate which, as near as possible, gives a grading approximating to the centre of the appropriate envelope shown. Adjustment of the mix proportions for subsequent trial mixes will then show whether some improvement in the quality of the mix is possible. The *Design of normal concrete mixes* published by the Building Research Establishment 1975 is a useful guide.

19.5 Workability of concrete and admixtures

Workability requirements for a concrete mix tend to conflict with requirements for maximum strength, density and economy, since workability increases with increased fines, cement, or water in a mix, but increased fines and water reduce density and strength, while increased cement may increase shrinkage and liability to cracking as well as adding to the cost of a mix. It is therefore necessary to produce minimum satisfactory workability in order to keep the deleterious effects of too much fines, cement or water to a minimum.

Workability can be measured by the well known **slump test**, but it is not very accurate and is best used only for ensuring a given mix is consistent,

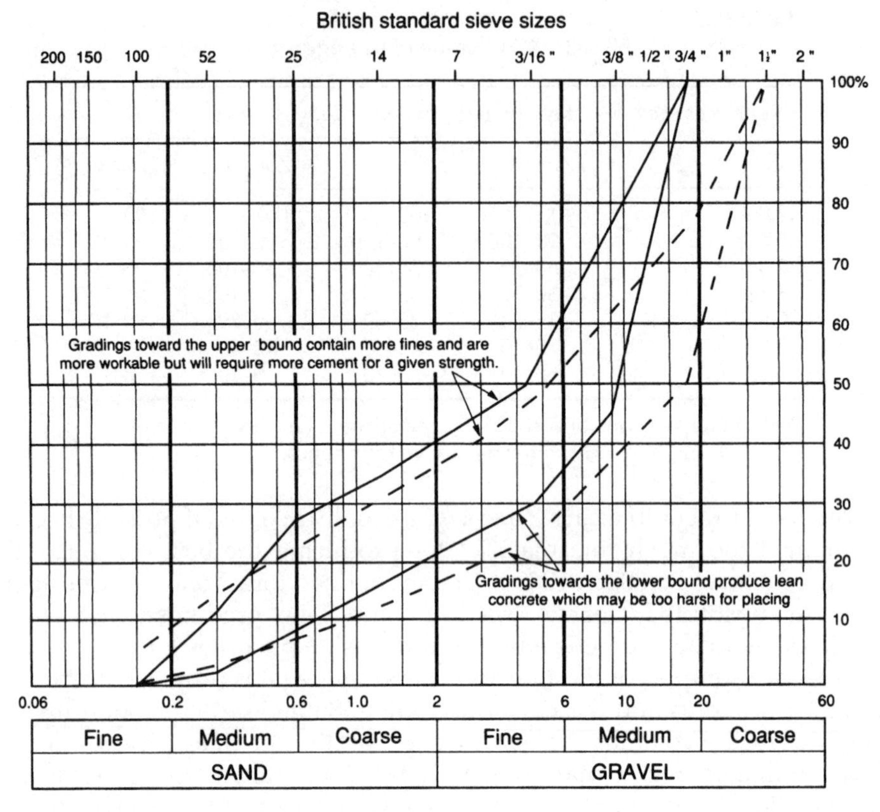

Fig. 19.3. Suitable range of aggregate gradings for concrete (based on modified versions of graphs by H.M. Walsh: *How to make good concrete*)

since slump varies with size and sharpness of aggregates used, as well as the amount of fines, cement and water in a mix. A truncated metal cone, 300 mm high by 100 mm diameter at the top and 200 mm at the bottom, is filled in three equal layers with concrete, each being rodded with 25 strokes of a 16 mm rod, rounded at both ends. On removal of the cone the 'slump' or drop in level of the top of the concrete below the 300 mm height is measured. Another site test uses the **compacting factor apparatus**, which works on the principle of finding the weight of concrete which falls via a sequence of two hoppers into a cylinder. The 'compacting factor' is the ratio of the weight of concrete falling into the cylinder as compared with the weight of concrete compacted to fill it. The higher this ratio is, the more workable is the concrete. These and other laboratory tests are described in BS 1881:1983. In practice, workability can be judged by eye as described in Section 19.6.

There is a substantial reduction of the workability of a concrete mix during the first 10 min after mixing, as anyone who has hand mixed concrete will know. This is primarily due to absorption of water by the aggregate so that the reduction in workability is less if the aggregate is wet before use. On a construction site this reduction of workability is not usually noticeable since more than 10 min usually passes before the concrete is placed. However, if samples for workability are taken, a time lapse of 10 min should be allowed before they are tested.

Admixtures to concrete are sometimes proposed by the contractor for approval, such as plasticizers or air-entraining agents to improve workability, or an accelerator to assist the contractor strike formwork early. If the specification does not define which admixtures or special cement can be used, the resident engineer should not agree to any such proposal from the contractor but pass it to the engineer for decision. There are many admixtures on the market, each having its own characteristics, some of which can be disadvantageous. For instance the use of the accelerator calcium chloride is not permitted for steel reinforced concrete under BS 8110 because it increases the risk of corrosion of the steel. Rapid hardening cement can cause a high concrete temperature leading to shrinkage and cracking; and air entraining agents reduce the density and strength of concrete. This does not mean that no admixtures should be permitted, but that the complex reactions they can cause make it necessary to call in specialist advice to ensure their safe use in any particular case. Very high strength concrete of up to 100 N/mm^2 strength or more, as used in high rise buildings, etc. is obtained primarily by use of a very low water/cement ratio. As a consequence adequate workability has to be achieved by use of an admixture. The specification must state precisely what is required.

Special aggregates for concrete usually comprise light-weight materials; mostly used only for particular building purposes (e.g. screeds for thermal or fire insulation, etc.), or used in precast concrete products.

19.6 Practical points in producing good concrete

Provided certain simple rules are followed good concrete can be achieved by methods varying from the 'bucket and spade' hand-labour method to use of the most sophisticated weigh-batching and mixing plant. The following shows the principal matters that should receive the resident engineer's attention.

First, choose **good aggregates**. The best guide is to use well-known local aggregates that have been and are being used satisfactorily on other jobs elsewhere. A reputable supplier will be able to name many jobs where his aggregate has been used, and the resident engineer will not be over-cautious if he visits one or two of these where the concrete is exposed to view. When the aggregates are being delivered on the job (not just the first few loads, but the loads when the supply has really got going), random loads as delivered should be examined. Handfuls of aggregate should be taken up and examined in detail, looking for small balls of clay, soft spongy stones, flaky stones, pieces of brick, soft shale, crumbly bits of sandstone, and whether clay or dirt is left on the hands after returning the handful. If the engineer finds more than one or two pieces of weak stone, or more than a single small piece of clay from a few handfuls, he should request the contractor to bring this to the notice of the supplier. He need not reject the load out of hand, but it will do no harm to let the supplier know the aggregates are being watched. If a load contains numerous weak stones or several pieces of clay, it should be rejected.

Diagnosing whether an aggregate is likely to give rise to alkali-silica reaction (which can cause expansion and disruption of concrete in a few years in the presence of moisture) requires specialist knowledge. The most practical approach for the engineer is to ask the supplier if his aggregate has been tested for this; if not, structures built some years previously with the aggregate should be checked for signs of cracking due to alkali-silica reaction. Guidance and precautions are set out in certain publications (References 1 and 2), but if it is proposed to use an aggregate not used before, the site staff should refer the problem to the engineer.

Second, choose **tested cement**. The same principle applies to cement as with the choice of aggregates; find the supplier of cement to other jobs and request a recent test certificate. Troubles can start when imported cement has to be used or cement from a variety of suppliers. Overseas it is not unusual for a small contractor to buy his cement a few bags at a time from the local bazaar. Testing such cement on site before any concrete is placed in an important part of a structure is essential. BS 12 provides methods for testing the compressive strengths of 1:3 mortar cubes or 1:2:4 concrete cubes but, if this is difficult to arrange, the flexural test mentioned in Section 19.3 can be applied on site.

Third, ensure **reasonably graded** aggregates. In delivery and stockpiling of coarse aggregate there is a tendency for the mix to segregate, the larger material remaining on top. Care has to be taken to ensure that certain batches are not made up from all the coarsest material and others from most of the fines. Crushed rock often has a considerable amount of dust in it, although this does not normally present a problem one does not want a batch made up mostly from dust and fines taken from the bottom of a stockpile.

Fourth, use **washed aggregates**. Unwashed aggregates suitable for concreting are rare: they are usually comprised of crushed clean homogeneous rock. Sometimes a river sand is supplied unwashed – it being assumed that the sand has already been 'washed' by the river. This should not be accepted as a fact, since a river also carries silts and clays. Sea-bed or beach sands must be washed in fresh water to remove the salt from them.

Fifth, achieve the **right workability**. Mechanical mixers are seldom at fault with regard to mixing, and hand-mixing can also be quite satisfactory; but it is the water content of a mix that requires the most vigilant attention. The site engineer should never let 'slop' be produced. Although the slump test and the compacting factor test are useful in defining the degree of stiffness of a mix, in practice judging the water content of a mix 'by eye' is both necessary and possible. The right sort of mix should look stiff as it comes out of the mixer or when turned over by hand on mixing boards. It should stand as a 'heap' and not as a 'pool' of concrete. When a shovel is thrust into such a pile, the shovel-cut should remain open for some minutes. Such a mix will look quite different after it is discharged and worked into some wet concrete already placed.

As soon as it is worked with shovels or vibrated, it will settle and appear to flow into and become part of the previously placed concrete.

The same characteristic makes it possible to judge the water content by noticing what happens if the freshly mixed concrete is carried in a dumper hopper to the point of discharge. The 'heap' of stiff concrete discharged from the mixer to the dumper hopper will appear to change to a pool of concrete as the dumper bumps its way round the usual site roads. When the dumper hopper is tipped, however, the concrete discharged should again appear stiff. But if, in transport, the concrete slops as a semi-fluid over the side of the dumper hopper, this shows too much water has been added. A simple density test on freshly mixed concrete (see Section 19.8) may assist in finding if the mix has too much water.

Sixth, **ram the concrete well in place**. Properly shovelled, rodded, or vibrated, the concrete should be seen to fill the corners of shuttering and to easily wrap around the reinforcing bars. When hand shovelling or rodding is adopted, it is scarcely possible to over-compact the concrete. But when mechanical vibrators are used the vibration should not be so prolonged as to produce a watery mix on the surface. Vibrators of the poker immersion type should be kept moving slowly in and out of the concrete. They should not be withdrawn quickly or they may leave an unfilled hole in the concrete; nor should they be left vibrating continuously in one location. Where vibrators are used, it is necessary for the contractor also to have available suitable hand rammers in case the vibrators break down in the middle of a pour.

Seventh, ensure the mix has **sufficient cement** in it. Normally contractors will use a little more cement than is theoretically necessary and this is helpful since batches of concrete vary. But if a contractor becomes too keen on cutting the cement to the bare minimum, a number of the cube crushing tests may fail to reach the required strength, and much delay may be caused by conducting the investigations required to seek out the cause.

19.7 Some causes of unsatisfactory concrete test results

The two most common kinds of failure are:

- failure to get the required strength, the concrete being otherwise apparently good;
- structural failures, such as honeycombing, sandy patches, and cracking.

Failure to get the right strength in cubes taken from a concrete pour can sometimes have a very simple cause. Among such causes are the following:

- the cube was not compacted properly;
- it was left out all night in hard frost or dried out in hot sun;
- there was a mix-up of cubes and a 7-day old cube was tested on the assumption it was 28 days old;
- the cube was taken from the wrong mix.

Such simple errors are not unusual and must be guarded against because they cause much perplexity and waste of time trying to discover the cause of a bad test result. The concrete must be fully compacted in the mould, which is kept under damp sacking until the next day when the mould can be removed and the cube marked for identity. It is then best stored in water at 'room temperature' for curing until sent to the test laboratory. If poor cube test results appear on consecutive batches, an error in the cement content of batches may be suspected, or else the quality of the cement itself.

Honeycombing is most usually caused by inadequate vibration or rodding of the concrete adjacent to the face of formwork. Sometimes too harsh a mix is used so there are insufficient fines to fill the trapped interstices between coarse aggregate and formwork, or the larger stones cause local arching. **Sand runs** – patches of sandy concrete on a wall surface which can be scraped away with a knife – can be due to over-vibration near a leaking joint in the formwork which allows cement and water to pass out of the mix. One simple, and not infrequent, cause of poor concrete is use of the wrong mix due to a 'failure of communication' with the batching plant operator or ready-mix supplier. An experienced concreting foreman should be able to detect a 'wrong mix' the moment it is discharged.

19.8 Site checks on concrete quality

The defect of cube and beam tests on concrete is that results cannot be known until some days after the concrete has been placed. If weak concrete appears to have been placed in a structure a difficult situation arises. The resident engineer can ask for the offending concrete to be demolished and re-built but this may pose such difficulty and delay that the decision ought not to be made on site without first discussing the problem with the engineer. The action taken depends upon how far the strength of the concrete falls short of the required strength, the load-bearing function of the under-strength concrete, and whether some alternative exists which does not involve breaking out the faulty concrete.

Frequent site checks of concrete quality can help to avoid such problems. Section 19.6 has already indicated that the water content of a mix can easily be judged by eye; and if the quality of the aggregate stocks held on site is kept under reasonable supervision, defects arising from aggregate quality or water content are unlikely to arise. Thus it is to the batching plant, and more particularly to the cement content, that checks should be directed.

One of the simplest on-the-spot tests which can be conducted is the **density** of freshly made concrete. This should be at least 2350 kg/m^3 (147 lb/ft^3) for a C20 mix and 2390 kg/m^3 (149 lb/ft^3) for a C30 mix on the assumption that the relative density of the aggregate is 2.65. The trial concrete mixes, however, should have revealed the typical densities expected for various grades of mix. The density can be obtained by filling and weighing an 0.015 m^3 (0.5 ft^3)

container with freshly mixed and compacted concrete. An adequately dense concrete cannot be made with badly graded aggregate or with an excess of water.

If mixing takes place on site the accuracy of the weigh-batching plant should be checked regularly. Actual errors found on a typical hand-operated weight batcher were:

- zero error on scale: up to 15 kg
- 20 mm stone: 78–106 per cent of required value
- sand: 97–125 per cent of required value
- cement (ex silo): 80–110 per cent of required value

Allowance in a mix has to be made for the weight of the **moisture content** of the sand which can be very variable when stocked in the open. Fig. 19.4 shows the relationship between the bulking factor and moisture content. Some devices are available for measuring the moisture content of a sand, but measuring the moisture in every batch is not a practical proposition. Instead, typical samples of sand from the stockpile under varying weather conditions can be weighed, then dried and weighed again. This gives a guide as to the weights of fine aggregate to be used under 'dry', 'moist', or 'wet' conditions. The moisture content of the coarse aggregate is not usually checked as it has little effect on the weight of the material.

Checking the **cement content** of the mix is particularly important if the cement is held in a silo. Serious under-weights of cement can occur due to machine faults with 'automatic' weighing equipment as well as with operator-controlled

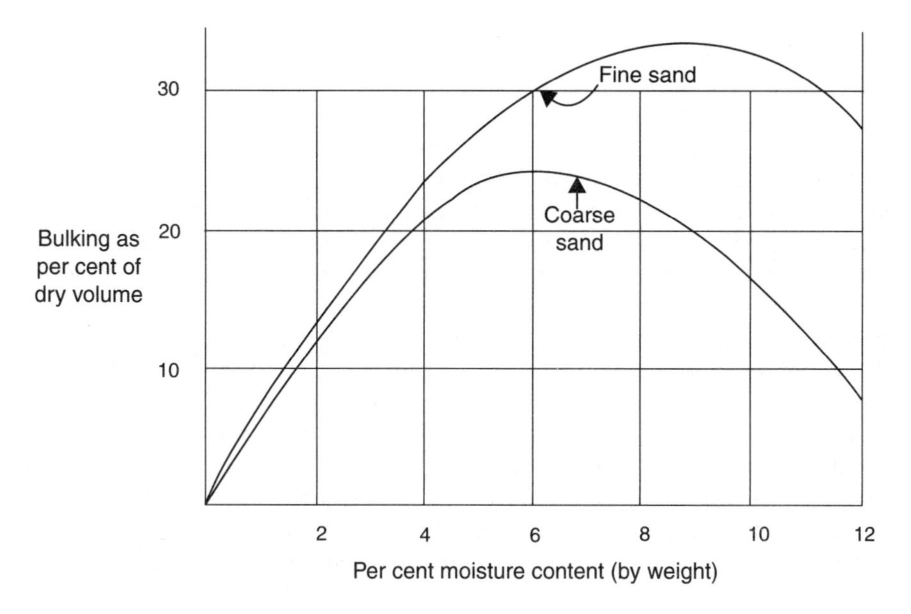

Fig. 19.4. Bulking of sand according to moisture content

discharges from the silo. It is better if concrete batches are made up per bag or (more usually) per 2 No. 50 kg bags of cement, in which case only variations in the weight of aggregate affect the mix; but this method is only possible for relatively modest concrete outputs, not when large pours are required. The cement content of a mix cannot be directly tested; hence the importance of keeping watch on the batching plant accuracy. It would not be unreasonable for the resident engineer to ask the contractor to conduct regular tests at suitable times on the accuracy of the batching plant. A responsible contractor will realize that it is better to ensure his plant is accurate, than to face the difficulty of finding that concrete placed is below the required strength.

Occasionally on small sites or overseas, **volume batching** of concrete is used. The weight per unit volume of aggregates has to be obtained by weighing the amount required to loosely fill a measured container. Suitable wooden gauge boxes for aggregate, sand and cement then have to be made up for a given mix. Average weights of Portland cement are 1280 kg/m^3 (80 lb/ft^3) loose, or 1440 kg/m^3 (90 lb/ft^3) when shaken. If hand mixing is adopted, fairly large gauge boxes with no bottom can be used, since they are placed on a mixing platform, filled and lifted off. They would usually be sized for 1 bag (50 kg) of cement. The bulking of the sand according to its moisture content has to be allowed for.

19.9 Conveyance and placing of concrete

Specifications often contain clauses dealing with the **transport of concrete**, requiring re-mixing after transport beyond a certain limit, limiting the height through which concrete can be dropped, and requiring no concrete be placed when more than a certain time has elapsed since mixing. In practice, problems of this sort seldom prove significant. Sometimes it may be necessary to insist that a contractor uses a closed chute to discharge concrete through a height in order to prevent segregation. Also it may be desirable to ensure mixed concrete is not left unplaced for over-long. A requirement often found in specifications is that concrete must not be placed after it reaches its 'initial set' which, for ordinary Portland cement concrete may take place 1–2 h after mixing, dependent on temperature, etc. However, a hardening on the outside due to surface drying can occur after about half-hour's standing, especially in hot weather. If this concrete is 'knocked up again' and shows it can be satisfactorily placed it need not be rejected. On the other hand, if a delay is so lengthy that the concrete hardens into lumps, such concrete must be discharged to waste.

Pumped concrete usually poses more problems for the contractor than it does for the resident engineer, since only well graded mixes relatively rich in cement are pumpable. Usually several mortar batches must be sent through the pipeline to 'lubricate it' before the first batch of concrete is pumped through, and pumping must thereafter be continuous. It is not easy to pump concrete more than 300–400 m. If a stoppage of the flow of concrete occurs for any reason, the contractor has to take swift action to prevent concrete solidifying in the pipeline. Compressed air is used to force the final concrete batch through the line, followed by water to clean the pipes. Plasticizers are frequently used in pumped concrete; these increase its workability without requiring increased cement or water. There are a wide variety based on different chemicals; BS 5075:1982 gives their main characteristics, but they should not be permitted by the resident engineer except to the extent allowed in the specification or sanctioned by the engineer.

Concrete can also be **blown** through a delivery pipe using a blower or compressed air. One batch at a time is blown through. The end of the delivery pipe must be directed into the area to be concreted, not against formwork which may be dislodged by the force of the ejected concrete. Proper warnings must be given to personnel before each 'shot' because aggregate can rebound and be dangerous, especially when blowing concrete into closed spaces such as the soffit to a tunnel lining.

The **skip method** of placing concrete is widely used. Skips can be either bottom-opening, or tip-over. In either case there can be a considerable bounce and sway of the skip when the concrete is discharged. The work should always be under the charge of an experienced ganger who keeps a continuous watch over the safety of his men.

19.10 Construction and other joints

The resident engineer must agree with the contractor where **construction joints** should be placed; but he should not require them to be placed in impracticable positions and must allow for the manner in which formwork must necessarily be erected. There are positions for construction joints which are 'traditional' even though the position may not seem to be the most desirable from a structural point of view. For instance a construction joint usually has to occur at the base of a wall even though it cantilevers from a base slab, which is a point of maximum tensile stress in one face of the wall concrete. This joint is best sited 150 mm above the base slab so as to give a firm fixing for the wall shutters and the best possibility of achieving a sound joint. In water-retaining work it is important to keep the number of construction joints to a minimum.

The **bonding** of one layer of concrete to a previous layer is usually accomplished by cleaning the surface of the old concrete with a high pressure water jet, and placing a layer at least 2 cm thick of mortar on the exposed surface immediately before the new concrete is placed. Sometimes a proprietary bonding mortar is used, especially when refilling cut-out portions of defective concrete. Wire brushing of the old surface is not so effective as water jetting, is laborious, and can seldom be properly done when reinforcement passes through a joint. A problem frequently encountered is that of finding debris on a construction joint at the bottom of erected formwork. Such debris must be removed before the mortar layer and new concrete is placed. Usually it is the job of the resident engineer's inspector to inspect formwork and the cleanliness of construction joints before permission is given to the contractor to start concreting. If the contractor runs 'Quality Assurance' one of his staff should act as inspector of formwork, but this does not relieve the resident engineer of his need to inspect on behalf of the engineer.

In **liquid-retaining structures** resilient plastic waterstops are usually provided at contraction joints. Fixing half their width in the stop-end shuttering to a narrow reinforced concrete wall often leaves a congested space for the concrete which must therefore be most carefully vibrated in place to ensure that the waterstop is bedded in sound concrete. If the concrete face of the joint is to be bitumen painted before the next wall section is built, bitumen must not get on the waterstop.

Floor joint grooves need cleaning out by water jetting, then surface drying as much as possible with an air blower before the priming compound supplied by the manufacturer of the joint filler is applied to the groove faces. It is essential that this primer is not omitted, and the filler must be pushed down to the bottom of the groove. Joint grooves are normally filled after the concrete has been allowed to dry out for 2 or 3 weeks when most shrinkage on drying should have taken place (see Section 19.11).

Leaks from liquid retaining concrete structures are most likely to occur from opening up of wall joints due to wall movement, especially at the corners of rectangular tanks; and puncturing of the floor joint filler under liquid pressure where the filler has not been solidly filled to the base of the groove.

19.11 Concrete finish problems

The skill required by carpenters to make and erect **formwork** for concrete is seldom fully appreciated. The formwork must remain 'true to line and level' despite substantial loading from the wet concrete. Column and wall faces have to be strictly vertical, and beam soffits strictly level, or any departure will be easily visible by eye. Formwork for concrete which is to remain exposed to view has to be planned and built as carefully as if it were a permanent feature of the building. Many methods have been tried to make the appearance of exposed concrete attractive: but any of them can be ruined by honeycombing, a bad construction joint, or by subsequent weathering revealing that one pour of concrete has not been identical with adjacent pours, or that the amount of vibration used in compacting one panel has been different from that used in others. If concrete has to remain exposed to public view, then the resident engineer should endeavour to agree with the contractor what is the most suitable method for achieving the finish required if the specification or drawings do not give exact guidance on the matter. The problem is that if, through lack of detailed attention, a 'mishap' on the exposed surface is revealed when the formwork is struck, it is virtually impossible to rectify it. Sometimes rendering the whole surface is the only acceptable remedy.

Where concrete will not remain exposed to view, minor discrepancies can be accepted. 'Fins' of concrete caused by the mix leaking through butt joints in the formwork should be knocked off. Shallow honeycombing should be chiselled out, and a chase cut along any defective construction joint. The cut-out area or chase should be washed, brushed with a thick cement grout, and then filled with a dryish mortar mix. This rectifying work should be done as soon as possible so the mortar mix has a better chance of bonding to the 'green' concrete.

Shrinkage cracking of concrete is a common experience. The shrinkage of concrete due to drying is of the order of 0.2–0.5 mm/m for the first 28 days. Subsequently concrete may expand slightly when wet and shrink on drying. The coefficient of temperature expansion or contraction is very much smaller, of the order of 0.007 mm/m per degree centigrade of change. Rich concrete mixtures tend to shrink more than lean mixes. The use of large aggregate, such as 40 mm instead of 20 mm, helps to minimize shrinkage. To avoid cracking of concrete due to shrinkage, wall lengths of concrete should be limited to about 9 m if restrained at the base or ends. Heavy foundations to a wall should not be allowed to stand and dry out for a long period before the wall is erected, because the wall concrete bonding to the base may be unable to shrink without cracking. Concrete is more elastic than is commonly appreciated, for example the unrestrained top of a 300 mm diameter reinforced concrete column 4 m high can be made to oscillate through nearly 1 cm by push of the hand.

19.12 Handling and fixing steel reinforcement

In best engineering practice the engineer will produce complete **bar-bending schedules** for use by the contractor. The engineer may not guarantee that such schedules are error free and may call upon the contractor to check them. But, as often as not, the contractor will fail to do this, so it is advisable for the resident engineer to check the schedules so that he can forewarn the contractor of any error present. In practice, few errors will be found because the advantage of producing bar-bending schedules is that it applies a detailed check on the validity of the reinforcement drawings supplied to the contractor.

In some contracts the contractor is required to produce bar-bending schedules himself from the reinforcement drawings supplied under the contract. This is not such good practice; the engineer foregoes an opportunity to check the reinforcement drawings, and the contractor (or his reinforcement supplier) who produces the bending schedules will not necessarily be sufficiently acquainted with the design to notice some discrepancy which indicates a possible design error.

Reinforcement is now seldom bent on site, except on sites overseas. Deliveries of reinforcement should be supervised by the leading steelfixer, who should check the steel against the bar schedules and direct where bars should be stocked. Bars should be delivered with identifying tags on them, but sometimes these get torn off. The leading steelfixer should not allow withdrawals from stock without his permission. If the contractor does not pay sufficient attention to this and, for example, lets various steelfixers pick what steel they think is right, the resident engineer should forewarn the contractor this is a recipe for ultimate chaos and delay.

Properly designed and bent bars can, in the hands of a good steelfixer, be as accurately placed as formwork. Crossings of reinforcement have to be wired together so that a rigid cage is built, able to withstand concrete placing without displacement. To ensure that the correct cover is given to bars, the contractor will need to prepare many small spacer blocks of concrete of the requisite cover thickness and about 25 mm square, which are wired onto the outside of reinforcement, keeping it the required distance from the formwork to give the specified cover. All wire ties should be snipped off close to the reinforcement so that their ends do not penetrate the concrete cover and form a path for corrosion of the reinforcement. The steelfixer will need to make and position spacer bars, generally U-shaped, which keep reinforcement layers the correct distance apart in slabs and walls. He may need many of these. They are not included in the bar-bending schedules and the cost to the contractor of supplying and fixing them is usually included in the price for steelfixing. Fig. 19.5 shows some points to watch when formwork and reinforcement is being erected.

Steel reinforcement stored on site rusts, but provided the rust is not so advanced that rust scales are formed, the rust does not appear to affect the bonding of the reinforcement to the concrete. A problem more likely to arise is the contamination of steel reinforcement with oil, grease, or bitumen. If the contractor wishes to oil or grease formwork to prevent it sticking to concrete, he should do so before the formwork is erected and not after it has been put in place. If the latter is attempted it will be almost impossible to prevent some oil or grease getting onto the reinforcement. Similarly, if contraction joints are to be bitumen painted, care must be taken not to get bitumen on bars passing through such a joint.

The proper **design and detailing** of reinforcement makes a major contribution to the quality and durability of reinforced concrete. The designer must choose diameters, spacings and lengths of bars which not only meet the theoretical design requirements but which make a practical system for erection and concreting. Reinforcement to slabs must either be strong enough for the steel fixer to stand on, or spaced far enough apart for him to get a foot between bars onto the formwork below. Wall and column reinforcement must be large enough diameter that it does not tend to sag under its own weight. Beam reinforcement should not be so congested that it will be difficult to get concrete to surround the bars without using a mix with too high a water content. The designer should consider options of design available to avoid heavy congestion of bars. An experienced designer who understands site erection problems will make as much use as possible of the four most commonly used bar diameters – 10, 12, 20 and 25 mm. He will appreciate that a 5 m long bar 25 mm diameter weighs about 20 kg, so that larger diameter or longer bars can be difficult for a steel fixer to handle on his own. For ease of handling, bars should not exceed 6–8 m length.

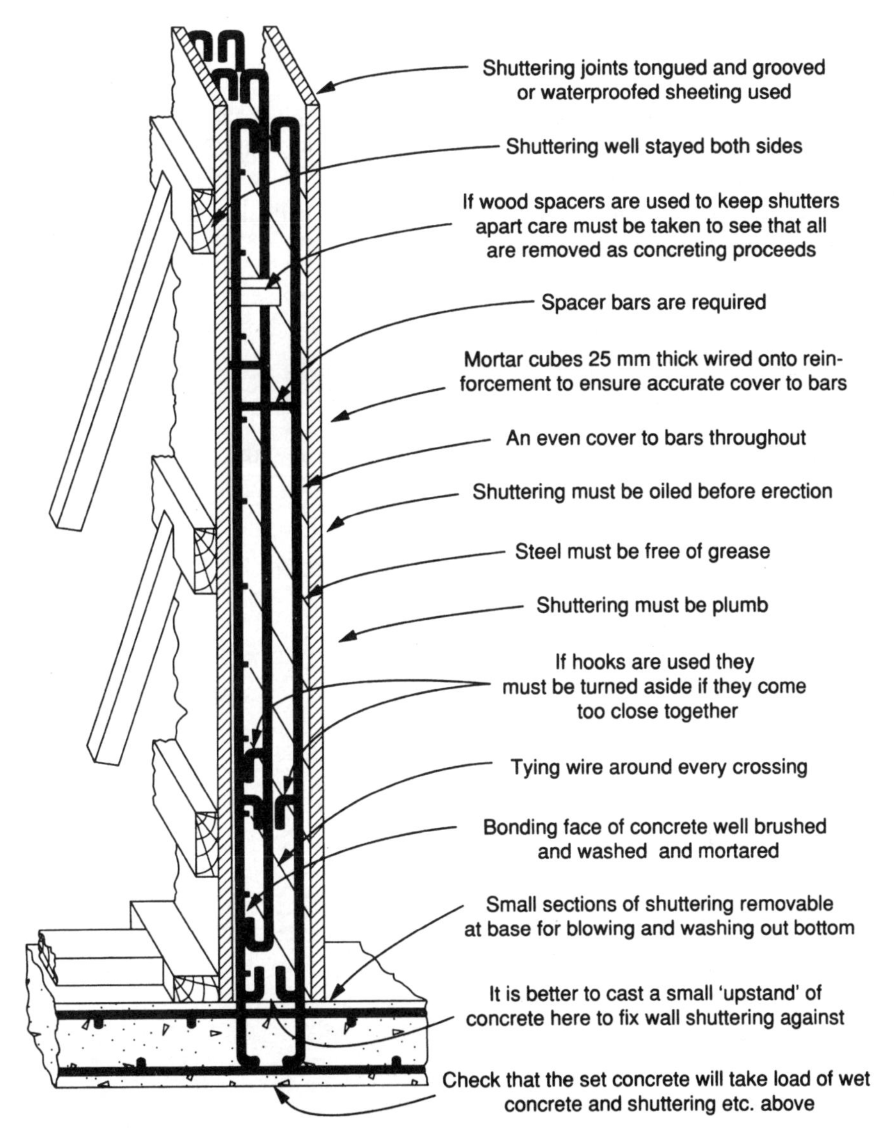

Fig. 19.5. Points to watch in erecting shuttering and reinforcement

Bond laps have to be allowed for and should be at places which are convenient for the erection of formwork and for concreting. Starter bars in floor slabs are nearly always necessary for bonding to the reinforcement in walls. The length of their vertical arm should not be longer than is necessary to provide adequate bond length and support the wall reinforcement so they present minimum impedance for slab concreting. If the designer wishes to use hooked bars, he should make sure that the thickness of slab or wall in which they are to be placed is sufficient to accommodate such hooks.

References

1. *Alkali–silica reaction: minimising the risk of damage to concrete. Guidance notes and model specification clauses*. Technical Report No. 30. Concrete Society, October 1987.
2. *Alkali aggregate reactions in concrete*. Digest 330. Building Research Establishment, March 1988.

Index